Electrical installation calculations

VOLUME 3

A. J. Watkins
B.Sc., B.Sc.(Aston), C.Eng., M.I.E.E.

THIRD EDITION
prepared by

Russell K. Parton
Formerly Head of the Department of Electrical and Motor Vehicle Engineering,
The Reid Kerr College, Paisley

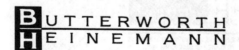
BUTTERWORTH
HEINEMANN

OXFORD   AUCKLAND   BOSTON   JOHANNESBURG   MELBOURNE   NEW DELHI

Butterworth-Heinemann
Linacre House, Jordan Hill, Oxford OX2 8DP
225 Wildwood Avenue, Woburn, MA 01801-2041
A division of Reed Educational and Professional Publishing Ltd

$\mathcal{R}$ A member of the Reed Elsevier plc group

First published by Edward Arnold 1957
Third edition 1999
Reprinted by Butterworth-Heinemann 2001

British Library Cataloguing in Publication Data
A catalogue record for this book is available from the British Library

ISBN 0 340 73186 9

Printed by St Edmundsbury Press Ltd, Bury St Edmunds, Suffolk

Contents

Preface to the third edition

This book together with Volumes 1 and 2 completes the series of three books intended for students of electrical installation work. They are essentially books of examples intended to co-ordinate the electrical installation technology, electrical science and the associated calculations of educational courses following the syllabuses of the City and Guilds and also electrical installation courses conducted under the auspices of BTEC and SCOTVEC. Volumes 1 and 2 are intended to satisfy the needs of City and Guilds Course 2360 Parts 1 and 2 and to lead to this Volume 3 which is devoted to the advanced level of course material for City and Guilds 2360 Course C.

Representative examples are worked and a selection of problems, with answers, provided to help the student to practise the answering techniques involved at final examination. Many of the examples are based on questions set in external examination papers.

In this, the third edition, the content has been revised taking into account the adoption by the British Standards Institution of the Institution of Electrical Engineers Wiring Regulations, sixteenth edition to form BS 7671: 1992.

The book revision also has regard to Amendments No 1, 1994 and No 2, 1997 to British Standard 7671. Readers are advised that a 'full' copy of BS 7671 with up-to-date amendments and also one of the 'site type' electricians' guides, e.g. the *IEE On-site Guide*, are essential when working through certain of the calculations.

Some foundation material from the previous two volumes is included where it was felt essential to enable readers to study the subject further.

The author and publishers are grateful to the British Standards Institution and to the Institution of Electrical

Engineers for permission to make use of data from their publications. Gratitude is also expressed to the City and Guilds for permission to use various questions from past examination papers; in certain cases slight rewording was necessary to satisfy modern terminology and requirements. The two Institutions and the City and Guilds however accept no responsibility for the interpretation by the author of their specific requirements or indeed for the answers to the various questions.

R.K.P.
Kilmacolm
1999

Basic circuit calculations

References: **Ohm's law**.   **Simultaneous equations**.
               Kirchhoff's laws.

WORKING STATEMENT OF KIRCHHOFF'S LAWS

1. The total current flowing towards any point in a circuit is equal to the total current flowing away from it.

    e.g. in Fig. 1,

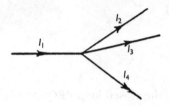

Fig. 1

$$I_1 = I_2 + I_3 + I_4$$

2. The sum of the voltage drops taken around a circuit is equal to the e.m.f. acting in the circuit.

    e.g. in Fig. 2,

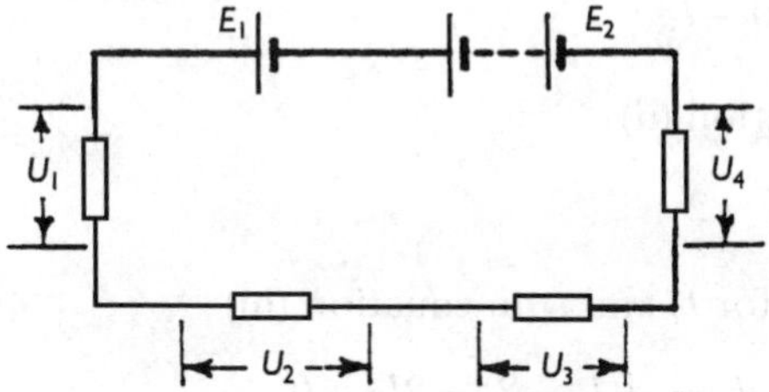

Fig. 2

$$E_1 + E_2 = U_1 + U_2 + U_3 + U_4$$

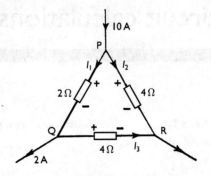

Fig. 3

EXAMPLE I Find the current in each resistor in Fig. 3.

Label the currents as shown in Fig. 3.

Applying Law 1 at P,

$$I_1 + I_2 = 10 \tag{i}$$

at Q $\quad I_3 + 2 = I_1$

or $\quad I_1 - I_3 = 2 \tag{ii}$

Applying Law 2 to the closed loop PRQ

$$4I_2 - 4I_3 - 2I_1 = 0 \tag{iii}$$

Note carefully the minus signs which allow for the fact that the voltage drops across I_1 and I_3 have polarities opposite to that across I_2.

From equation (i)

$$I_2 = 10 - I_1 \tag{iv}$$

From equation (ii)

$$I_3 = I_1 - 2 \tag{v}$$

Substituting for I_2 and I_3 in equation (iii)

$$4(10 - I_1) - 4(I_1 - 2) - 2I_1 = 0$$
$$40 - 4I_1 - 4I_1 + 8 - 2I_1 = 0$$
$$48 - 10I_1 = 0$$

or $\qquad 48 = 10I_1$

$$\therefore \quad I_1 = \frac{48}{10}$$

$$= \underline{4.8\,\text{A}}$$

Substituting in equation (iv) $\qquad I_2 = 10 - 4.8$

$$= \underline{5.2\,\text{A}}$$

Substituting in equation (v) $\qquad I_3 = 4.8 - 2$

$$= \underline{2.8\,\text{A}}$$

Check at R where $\qquad\qquad I_2 + I_3 = 8$

$$5.2 + 2.8 = 8$$

Currents determined for each part of the circuit have the values shown above and directions as indicated on the circuit diagram.

EXAMPLE 2 Write down a set of simultaneous equations for the network shown in Fig. 4.

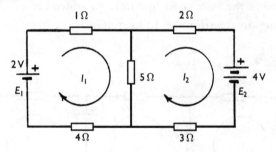

Fig. 4

Imagine independent currents I_1 and I_2 to circulate in a *clockwise* direction as shown. The part of the circuit carrying the current I_1 is called Mesh 1, that part carrying the current I_2 is called Mesh 2. An e.m.f. is said to be positive if it acts in the same direction as the assumed current. For example E_1 is positive, E_2 is negative.

Applying the second Kirchhoff law and remembering that

$$U = I \times R$$

Mesh 1 $\quad E_1 = 1I_1 + 5I_1 + 4I_1 - 5I_2$

3

The last term allows for the voltage drop produced in the $5\,\Omega$ resistor by the current I_2 which passes through the resistor in the opposite direction to that of I_1. This equation simplifies to

$$2 = 10I_1 - 5I_2$$

Similarly for Mesh 2

$$-E_2 = 2I_2 + 3I_2 + 5I_2 - 5I_1$$

or $\qquad -4 = -5I_1 + 10I_2$

Notice that the current terms have been rearranged so that they appear in number order from left to right. The two equations are now written together and labelled

$$2 = 10I_1 - 5I_2 \tag{1}$$
$$-4 = -5I_1 + 10I_2 \tag{2}$$

The following examples (3–11) illustrate more complicated circuits. The student will find it helpful to examine the circuits and to follow the associated equations provided before attempting the questions set in Exercise 1.

EXAMPLE 3

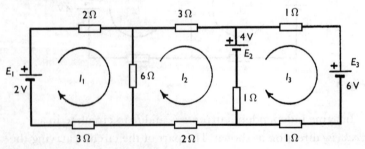

Fig. 5

$E_1 = 2\,\text{V}$ positive sign

$E_2 = 4\,\text{V}$ negative sign for I_2 positive sign for I_3

$E_3 = 6\,\text{V}$ negative sign.

Mesh 1: $\qquad 2 = 2I_1 + 6I_1 + 3I_1 - 6I_2$

or $\qquad 2 = 11I_1 - 6I_2$

Mesh 2: $-4 = 3I_2 + 1I_2 + 2I_2 + 6I_2 - 1I_3 - 6I_1$

or $-4 = -6I_1 + 12I_2 - 1I_3$

Mesh 3: $4 - 6 = 1I_3 + 1I_3 + 1I_3 - 1I_2$

or $-2 = -1I_2 + 3I_3$

and writing the three equations together

$$2 = 11I_1 - 6I_2 \tag{1}$$

$$-4 = -6I_1 + 12I_2 - 1I_3 \tag{2}$$

$$-2 = -1I_2 + 3I_3 \tag{3}$$

EXAMPLE 4

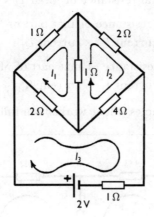

Fig. 6

Figure 6 is a Wheatstone Bridge.

$$E_1 = 0$$

$$E_2 = 0$$

$$E_3 = 2\,\text{V positive sign.}$$

Mesh 1: $0 = 1I_1 + 1I_1 + 2I_1 - 1I_2 - 2I_3$

or $0 = 4I_1 - 1I_2 - 2I_3$

Mesh 2: $0 = 2I_2 + 4I_2 + 1I_2 - 1I_1 - 4I_3$

or $0 = -1I_1 + 7I_2 - 4I_3$

Mesh 3:　　$2 = 2I_3 + 4I_3 + 1I_3 - 2I_1 - 4I_2$

or　　　　$2 = -2I_1 - 4I_2 + 7I_3$

and writing the equations together

$$0 = 4I_1 - 1I_2 - 2I_3 \tag{1}$$

$$0 = -1I_1 + 7I_2 - 4I_3 \tag{2}$$

$$2 = -2I_1 - 4I_2 + 7I_3 \tag{3}$$

It is useful to note (i) that in this form, all the terms on the right-hand side are negative except for a diagonal row of positive terms; (ii) the equations may be remembered in words, for example in the case of Mesh 1:

e.m.f. = [(total resistance of Mesh 1) × current I_1]

　　　　− [(resistance of that part of Mesh 1 carrying current I_2) × current I_2]

　　　　−[(resistance of that part of Mesh 1 carrying current I_3) × current I_3].

The following examples will be worked out fully.

EXAMPLE 5

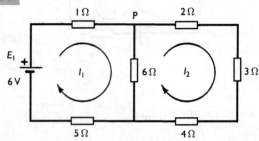

Fig. 7

Calculate the p.d. across the 6-ohm resistor in Fig 7.

　　　$E_1 = 6\,\text{V positive sign}$

　　　$E_2 = 0$

Mesh 1:　$6 = 1I_1 + 6I_1 + 5I_1 - 6I_2$

Mesh 2:　$0 = 2I_2 + 3I_2 + 4I_2 + 6I_2 - 6I_1$

which are simplified and rewritten as

$$6 = 12I_1 - 6I_2 \tag{1}$$

$$0 = -6I_1 + 15I_2 \tag{2}$$

From equation (2)

$$6I_1 = 15I_2$$

or $$I_1 = \frac{15I_2}{6} \tag{3}$$

Substituting in equation (1)

$$6 = 12 \times \frac{15I_2}{6} - 6I_2$$

or $$6 = 30I_2 - 6I_2$$

or $$6 = 24I_2$$

$$\therefore \quad I_2 = \frac{6}{24} = \underline{0.25\,\text{A}}$$

Substituting in equation (3)

$$I_1 = \frac{15}{6} \times \frac{6}{24} = \underline{0.625\,\text{A}}$$

The fact that both currents have positive signs means that the actual current flows in the same direction as the assumed current. The current in the 6-ohm resistor is

$$I_1 - I_2 = 0.625 - 0.25$$

$$= 0.375\,\text{A}$$

Take care to insert the correct signs for I_1 and I_2.

The polarity of the battery ensures that this current flows from P to Q. The p.d. across the 6-ohm resistor is

$$U_6 = 0.375 \times 6$$

$$= \underline{2.25\,\text{V}}$$

P being at the higher potential (or P is positive with respect to Q).

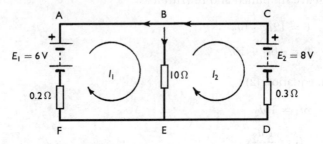

Fig. 8

EXAMPLE 7

Find the current through each part of the circuit of Fig. 8.

E_1 is positive with respect to I_1

E_2 is negative with respect to I_2

Mesh 1: $6 = 10.2I_1 - 10I_2$ (1)

Mesh 2: $-8 = -10I_1 + 10.3I_2$ (2)

From (1) $I_2 = \dfrac{10.2I_1 - 6}{10}$

$= 1.02I_1 - 0.6$ (3)

Substitute in (2) $-8 = -10I_1 + 10.3(1.02I_1 - 0.6)$

$-8 = -10I_1 + 10.5I_1 - 6.18$

$-8 = 0.5I_1 - 6.18$

$I_1 = \dfrac{6.18 - 8}{0.5}$

$I_1 = \underline{-3.64\,\text{A}}$

This is the actual current in the parts of the circuit BA, AF and FE. The minus sign indicates that the actual current flows in the reverse direction to the assumed current I_1. We shall indicate the actual current as $I_{BA} = I_{AF} = I_{FE} = 3.64\,\text{A}$; the order of the letters then indicates the direction of the actual current. Substitute for I_1 in (3)

$I_2 = 1.02(-3.64) - 0.6$

$= -3.71 - 0.6$

$I_2 = \underline{-4.31\,\text{A}}$

The minus sign again shows that the actual current flow is opposite to that of I_2, that is

$$I_{ED} = I_{DC} = I_{CB} = 4.31\,A$$

As an alternative to the method of example E, mark on the diagram say at B the actual directions of current flow using small arrows as shown. The current in the 10-ohm resistor is then found by applying Kirchhoff's first law at point B;

thus current in 10-ohm resistor

$$= I_{BE} = I_{CB} - I_{BA}$$

or $\quad = 4.31 - 3.64$

$\quad = \underline{0.67\,A}$ in direction from B to E.

(The current must flow in this direction because both batteries have their positive terminals at the top.)

EXAMPLE 8 Two batteries connected in parallel are to be charged by connecting them in series with a $0.8\,\Omega$ resistor to a 20 V d.c. supply. Battery 1 has e.m.f. 12 V and internal resistance $0.2\,\Omega$. Battery 2 has e.m.f. 12 V and internal resistance $0.4\,\Omega$. Calculate the current taken from the supply and the current through each battery.

The circuit arrangement is shown in Fig. 9.

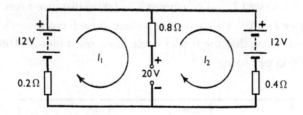

Fig. 9

Mesh 1: $\quad 12 - 20 = 1I_1 - 0.8I_2$

Mesh 2: $\quad 20 - 12 = -0.8I_1 + 1.2I_2$

or $\quad\quad\quad\quad -8 = I_1 - 0.8I_2$ $\hfill (1)$

$\quad\quad\quad\quad\quad\quad 8 = -0.8I_1 + 1.2I_2$ $\hfill (2)$

From (1) $\qquad I_1 = 0.8I_2 - 8$ $\hfill(3)$

Substituting for I_1 in (2)

$$8 = -0.8(0.8I_2 - 8) + 1.2I_2$$
$$8 = -0.64I_2 + 6.4 + 1.2I_2$$
$$8 - 6.4 = 0.56I_2$$
$$I_2 = \frac{1.6}{0.56}$$
$$= 2.85 \, \text{A}$$

Thus battery 2 is being charged at the rate of 2.85 A. Substituting for I_2 in (3)

$$I_1 = 0.8 \times \frac{1.6}{0.56} - 8$$
$$= -5.714 \, \text{A}$$

Thus battery 1 is being charged at the rate of 5.714 A. The supply current is thus

$$5.714 + 2.85 = \underline{8.564 \, \text{A}}$$

EXAMPLE 9   A substation S feeds a d.c. distributor at 235 V. Consumer X at 200 m distance takes 120 A, consumer Y at 450 m distance takes 80 A and consumer Z at 600 m distance takes 100 A (see Fig. 10). Calculate the voltage at each consumer's terminals given that the feeder cable has a resistance of 0.08 Ω per 1000 m per core.

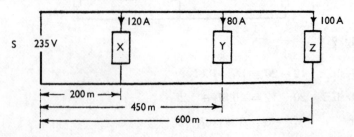

Fig. 10

Applying the first Kirchhoff law:

Current in section YZ $= 100\,\text{A}$

Current in section XY $= 100 + 80 = 180\,\text{A}$

Current in section SX $= 180 + 120 = 300\,\text{A}$

Resistance of 1 m of *double* core $= \dfrac{0.8}{1000} \times 2\,\Omega$

$$\begin{aligned}
\text{Resistance of section SX} \quad &= 200 \times \frac{0.08}{1000} \times 2 \\
&= 0.032\,\Omega
\end{aligned}$$

$$\begin{aligned}
\text{Voltage drop in SX} \quad &= 300 \times 0.032 \\
&= 9.6\,\text{V}
\end{aligned}$$

$$\begin{aligned}
\therefore \quad \text{voltage at X terminals} \quad &= 235 - 9.6 \\
&= 225.4\,\text{V}
\end{aligned}$$

$$\begin{aligned}
\text{Resistance of section XY} &= 250 \times \frac{0.08}{1000} \times 2 \\
&= 0.04\,\Omega
\end{aligned}$$

$$\begin{aligned}
\text{Volts drop in section XY} &= 180 \times 0.04 \\
&= 7.2\,\text{V}
\end{aligned}$$

$$\begin{aligned}
\text{voltage at Y terminals} \quad &= 225.4 - 7.2 \\
&= 218.2\,\text{V}
\end{aligned}$$

$$\begin{aligned}
\text{Resistance of section YZ} &= 150 \times \frac{0.08}{1000} \times 2 \\
&= 0.024\,\Omega
\end{aligned}$$

$$\begin{aligned}
\text{Volts drop in section YZ} &= 100 \times 0.024 \\
&= 2.4\,\text{V}
\end{aligned}$$

$$\begin{aligned}
\text{voltage at Z terminals} \quad &= 218.2 - 2.4 \\
&= 215.8\,\text{V}
\end{aligned}$$

Correct to three significant figures the required voltages are
thus – at X 225 V

  at Y 218 V

  at Z 216 V

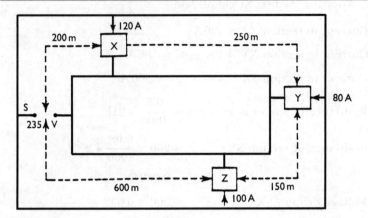

Fig. 11

EXAMPLE 10 If the distributor of Example 9 be reconnected so as to form a ring circuit (see Fig. 11), calculate the current in each section and the voltage at each consumer's terminals.

Applying the first Kirchhoff law and calling the current in section SX xA

then the current in section XY $= (x - 120)$ A

and current in section YZ $\quad = (x - 120 - 80)$

$\qquad = (x - 200)$ A

and current in section ZS $\quad = (x - 200 - 100)$

$\qquad = (x - 300)$ A

As before, resistance of SX $\quad = 0.032\,\Omega$

$\therefore$ volts drop in section SX $\quad = 0.032x\,$V

resistance of XY $\quad = 0.04\,\Omega$

$\therefore$ volts drop in section XY $\quad = 0.04(x - 120)\,$V

resistance of YZ $\quad = 0.024\,\Omega$

$\therefore$ volts drop in section YZ $\quad = 0.024(x - 200)\,$V

resistance of section ZS $\quad = 600 \times \dfrac{0.08 \times 2}{1000}$

$\qquad = 0.096\,\Omega$

volts drop in section ZS $\quad = 0.096(x - 300)\,$V

There is no resultant e.m.f. round the ring circuit formed by either conductor since both start and finish at S. Thus

$$0.032x + 0.04(x - 120) + 0.024(x - 200) + 0.096(x - 300) = 0$$

$$0.032x + 0.04x - 4.8 + 0.024x - 4.8 + 0.096x - 28.8 = 0$$

$$0.192x = 38.4$$

$$x = \frac{38.4}{0.192}$$

$$= 200 \, \text{A}$$

i.e. the current in section SX = 200 A

Volts drop in section SX	$= 200 \times 0.032$
	$= 6.4 \, \text{V}$
so that the voltage at X	$= 235 - 6.4$
	$= 228.6 \, \text{V}$
Current in section XY	$= x - 120$
	$= 200 - 120 = 80 \, \text{A}$
Volts drop in section XY	$= 80 \times 0.04$
	$= 3.2 \, \text{V}$
so that the voltage at Y	$= 228.6 - 3.2$
	$= 225.4 \, \text{V}$
Current in section YZ	$= x - 200 \, \text{A}$
	$= 200 - 200$
	$= 0$
The volts drop in YZ	$= 0$
and voltage at Z	$= 225.4 \, \text{V}$
current in ZS	$= (x - 300)$
	$= 200 - 300$
	$= -100 \, \text{A}$ (flowing in opposite direction)
Volts drop in section ZS	$= -100 \times 0.096$
	$= -9.6 \, \text{V}$ (a rise in voltage)
and the voltage at S	$= 225.4 - (-9.6)$
	$= 235.0$

To summarise, and correcting to three significant figures.

Current in SX = 200 A, p.d. at X = 229 V

Current in XY = 80 A, p.d. at Y = 225 V

Current in YZ = 0, p.d. at Z = 225 V

Current in ZS = 100 A

EXAMPLE 11 A two-wire distribution cable PQRST, 200 m long and of uniform cross-sectional area (see Fig. 12), is supplied at end P at 250 V, 50 Hz single phase, and at end T at 247 V, 50 Hz single phase. Steady loads are connected to the cable as follows: 40 A at Q, 60 A at R and 50 A at S, where Q, R and S are respectively 50 m, 100 m and 160 m from P. The resistance of the complete cable (lead and return) is 0.2 Ω. Cable reactance may be ignored.

(a) Find the values of the currents entering the cable at P and T.

(b) Find the potential difference at R. (C & G)

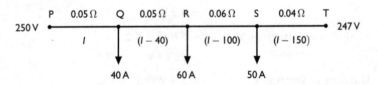

Fig. 12

Total volts drop between P and T = 250 − 247 = 3 V

Cable resistance PQ $\dfrac{50}{200} \times 0.2 = 0.05\,\Omega$

QR $\dfrac{50}{200} \times 0.2 = 0.05\,\Omega$

RS $\dfrac{60}{200} \times 0.2 = 0.06\,\Omega$

ST $\dfrac{40}{200} \times 0.2 = 0.04\,\Omega$

Let I be the current in section PQ entering at P.

Then current in section:

$$QR = (I - 40)\,A$$
$$RS = (I - 100)\,A$$
$$ST = (I - 150)\,A$$

Volts drop in section:

$$PQ = 0.5 \times I$$
$$QR = 0.05 \times (I - 40)$$
$$RS = 0.06 \times (I - 100)$$
$$ST = 0.04 \times (I - 150)$$

Adding volts drops

$$0.05 \times I + 0.05 \times (I - 40) + 0.06 \times (I - 100)$$
$$+ 0.04 \times (I - 150) = 250 - 247$$

So
$$0.2I - 14 = 3$$
$$0.2I = 17$$
$$I = \frac{17}{0.2} = 85\,A$$

(a) Current entering at $P = I = 85\,A$

Current entering at $T = (I - 150) = (85 - 150)$
$$= 65\,A$$

(b) Potential at $R = 250 - (\text{volt drop } PQ + \text{volt drop } QR)$
$$= 250 - (0.05 \times 85 + 0.05 \times 45)$$
$$= 250 - 6.5$$
$$= 243.5\,V$$

EXERCISE I

1. Write down sets of simultaneous equations for the circuits shown in Fig. 13. (Resistance values are all in ohms.)

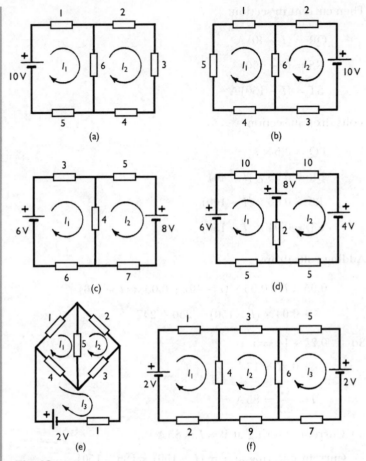

Fig. 13

2. Find the current in each part of each circuit shown in Fig. 14.
   (Resistance values are all in ohms.)

3. Find the current in section BD of the bridge circuit shown in Fig. 15.

4. Two batteries are connected in parallel and together they supply
   current to a $10\,\Omega$ resistor connected across their terminals. One
   battery has e.m.f. $6\,V$ and internal resistance $1\,\Omega$; the other has e.m.f. $6\,V$
   and internal resistance $2\,\Omega$. Calculate the current through the resistor
   and the p.d. between its ends.

5. The circuit of question 4 is rearranged so that the batteries are charged
   from a 10-volt supply in series with the 10-ohm resistor. Calculate the
   current through each battery and the total charging current.

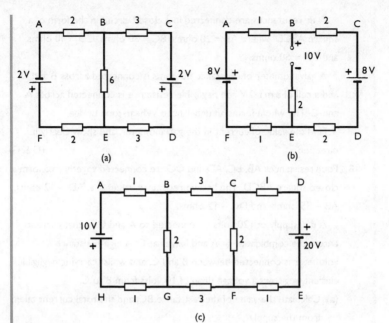

(a)

(b)

(c)

Fig. 14

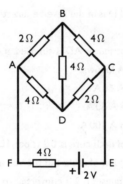

Fig. 15

6. A battery of cells of total e.m.f. 40 volts and with total internal resistance 2 ohms, is connected in parallel with a second battery of 44 volts and internal resistance 4 ohms. A load resistance of 6 ohms is connected across the ends of the parallel circuit.

   Calculate the currents in each battery and in the load resistance. Draw a diagram and show the current directions. (C & G)

7. Write down Kirchhoff's laws, and use them in working the following calculation.

Four resistances are connected in a closed circuit in the form of a square ABCD, where AB = 20 ohms, BC = 30 ohms, CD = 40 ohms and DA = 50 ohms.

A galvanometer of resistance 80 ohms is connected across B and D, and a cell of e.m.f. 2 V and negligible resistance is connected across A and C. The whole forms an unbalanced Wheatstone Bridge.

Find the value of current in the galvanometer, and show clearly its direction. (C & G)

8. Four resistances AB, BC, AD and DC are connected together to form a closed square ABCD. The known resistance values are: AD – 12 ohms, AB – 35 ohms, and DC – 12 ohms.

A d.c. supply of 120 volts is connected to A and C so that current enters the combination at A and leaves at C. A high-resistance voltmeter is connected between B and C, and whilst carrying negligible current, registers a voltage drop of 10 volts from B to C.

(a) Calculate the value of the resistance BC, and the total current taken from the supply.

(b) Calculate also the value of BC, such that the potential difference between B and D is in the reverse direction, i.e. from D to B.

(C & G)

9. A d.c. two-wire distributor AD is supplied at end A at 230 V and loads are taken from it as follows:

at B 50 m from A 60 A
at C 200 m from A 80 A
at D 300 m from A 100 A.

The resistance of each core is 0.2 Ω per 1000 m. Calculate the current in each section of cable and the voltage at B, C and D.

10. The distributor of question 9 is converted to a ring circuit by joining D to A with a 300 m length of the same cable. Calculate the current in each section and the voltage at B, C and D.

11. Write down Kirchhoff's laws and use them in working the following calculation:

A two-wire ring main is 600 m long, and is supplied at a point P with direct current at 240 V. A load of 70 A is taken from the main at point Q, 150 m from P in one direction, and a further load of 50 A is taken at a point R, 300 m from P in the other direction. The resistance of the main is 0.4 Ω/1000 m of single core.

Find the currents in value and direction in each section of the ring main, and calculate the voltages at points Q and R.   (C & G)

12. A three-wire d.c. system supplies power to two adjacent workshops by means of a three-core aluminium cable. The voltage at the main switchboard is maintained at 240/0/240 V.

Workshop A takes 200 A from positive and neutral, and workshop B takes 250 A from neutral and negative. The resistance of each outer core of the cable is 0.238 Ω and that of the inner core is 0.396 Ω.

Show on a suitable diagram the values and directions of the currents in the cable, and calculate the voltages at A and B respectively.

Describe in general terms what would happen if the neutral core became disconnected.

Alternating current circuit calculations I
Series circuits

References: **Inductive reactance.**   **Phasor diagrams.**
           Capacitive reactance.   **Quadratic equations.**
           Impedance.
           Power and power factor.
           Resonance.

Consider a circuit consisting of resistance, inductance and capacitance in series (see Fig. 16).

The current is the same at all places in the circuit.

The voltage drop across the resistance is $U_R = I \times R$ in phase with the current.

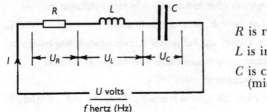

Fig. 16

R is resistance (ohm)
L is inductance (henry)
C is capacitance (microfarad)

The voltage drop across the inductance is $U_L = I \times X_L$ where $X_L = 2\pi fL$. This voltage drop leads on the current by $90°$.

The voltage drop across the capacitor is $U_C = I \times X_C$ where

$$X_C = \frac{10^6}{2\pi fC}$$

This voltage drop lags on the current by $90°$.

The supply voltage is U.

The phasor diagram is usually shown as in Fig. 17.

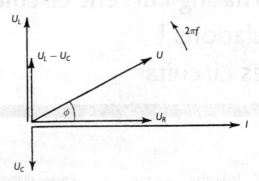

Fig. 17

The current is drawn first as the reference phasor. The resultant of U_L and U_C, since they are in antiphase, is found by subtracting the smaller from the larger. The supply voltage U is drawn by combining U_R and $U_L - U_C$ (or $U_C - U_L$ in the manner described in Volume 2).

ϕ is the phase angle between current and supply voltage. Cos ϕ is the overall power factor of the circuit. Figure 17 shows the phasor diagram for an overall lagging power factor. This occurs when U_L is greater than U_C.

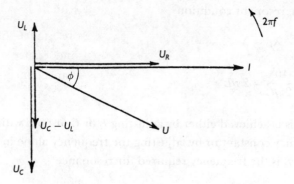

Fig. 18

Figure 18 shows the case of an overall leading power factor which occurs when U_C is greater than U_L.

Figure 19 shows the case when $U_L = U_C$; here $U = U_R$.

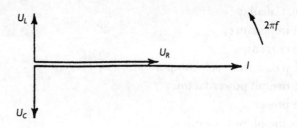

Fig. 19

The phase angle between current and supply voltage is zero giving a power factor of 1 or unity ($\because \cos 0 = 1$). This is the series resonant condition; the current in the circuit has its maximum value, being limited only by the resistance.

The impedance of the general series circuit is given by

$$Z^2 = R^2 + (X_L - X_C)^2$$

or $\quad Z^2 = R^2 + (X_C - X_L)^2$

(simply taking the smaller from the larger) and from either formula it is seen that when $X_C = X_L$

$$Z = R$$

For the resonant condition

$$X_C = X_L$$

$$\frac{10^6}{2\pi fC} = 2\pi fL$$

and this is achieved either by adjusting L or C or both with the frequency constant or by adjusting the frequency alone in which case if f_r is the frequency required for resonance

$$f_r = \frac{10^3}{2\pi\sqrt{LC}} \text{ Hz}$$

EXAMPLE I   A $10\,\Omega$ resistor, a $100\,\mu\text{F}$ capacitor and an inductor of $0.15\,\text{H}$ are connected in series to a supply at $230\,\text{V}$ $50\,\text{Hz}$ (see Fig. 20). Calculate:
(a) the impedance;
(b) the current;
(c) the p.d. across each component;
(d) the overall power factor;
(e) the power.
   Draw the phasor diagram.

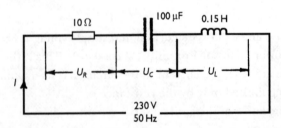

Fig. 20

$$X_L = 2\pi fL \qquad\qquad X_C = \frac{10^6}{2\pi fC}$$

$$= 2\pi \times 50 \times 0.15$$

$$= 47.1\,\Omega \qquad\qquad\qquad = \frac{10^6}{2\pi \times 50 \times 100}$$

$$= 31.83\,\Omega$$

$$Z^2 = R^2 + (X_L - X_C)^2$$

$$= 10^2 + (47.1 - 31.83)^2$$

$$= 10^2 + (15.3)^2$$

$$= 100 + 234.2$$

$$Z = \sqrt{334.2}$$

$$= \underline{18.28\,\Omega} \qquad \text{(a)}$$

$$I = \frac{U}{Z}$$

$$= \frac{230}{18.28}$$

$$= \underline{12.58\,\text{A}} \qquad \text{(b)}$$

$$U_R = I \times R$$

$$= 12.58 \times 10$$

$$= \underline{125.8\,\text{V}}$$

$$U_C = I \times X_C$$

$$= 12.58 \times 31.83$$

$$= \underline{400\,\text{V}}$$

$$U_L = I \times X_L$$

$$= 12.58 \times 47.1$$

$$= \underline{592.5\,\text{V}} \qquad \text{(c)}$$

From the phasor diagram (Fig. 21)

$$\tan\phi = \frac{U_L - U_C}{U_R}$$

$$= \frac{592.5 - 400}{125.8}$$

$$= \frac{192.5}{125.8}$$

$$= 1.53$$

$$\phi = 57°$$

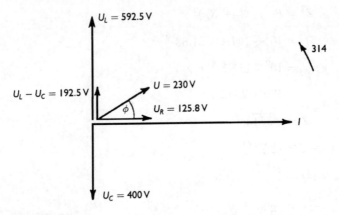

Fig. 21 Phasor diagram

Power factor $= \cos\phi$ (lag)

$$= 0.545 \qquad \text{(d)}$$

Total power

$$P = UI\cos\phi$$
$$= 230 \times 12.58 \times 0.545$$
$$= \underline{1576.9\,\text{W}} \qquad \text{(e)}$$

EXAMPLE 2   Calculate the value to which the frequency of the previous circuit would have to be adjusted in order for the power factor to be unity and determine the current at this frequency.

Unity power factor is achieved at the resonant condition. The required frequency is

$$f_r = \frac{10^3}{2\pi\sqrt{LC}}$$
$$= \frac{10^3}{2\pi\sqrt{0.15 \times 100}}$$
$$= \frac{10^3}{2\pi \times \sqrt{15}}$$
$$= \frac{10^3}{2\pi \times 3.873}$$
$$= \underline{41.1\,\text{Hz}}$$

At this frequency the two values of reactance are equal and the current is limited only by the resistance. The current is then

$$I_r = \frac{U}{R}$$

$$= \frac{230}{10}$$

$$= \underline{23\,\text{A}}$$

EXAMPLE 3 A coil has resistance $80\,\Omega$ and inductance $0.318\,\text{H}$. Calculate the value of a capacitor which if connected in series with the coil to a $50\,\text{Hz}$ sinusoidal supply will cause the same current to flow as if the coil alone were connected to the same supply. Draw the phasor diagrams and determine the overall power factor.

Reactance of coil

$$X_L = 2\pi f L$$

$$= 2\pi \times 50 \times 0.318$$

$$= 100\,\Omega$$

Impedance of coil alone

$$Z^2 = R^2 + X_L^2$$

$$= 80^2 + 100^2$$

$$Z = 128\,\Omega$$

Impedance of the coil and capacitor in series

$$Z^2 = R^2 + (X_C - X_L)^2$$

and for the same current these two values must be the same, therefore

$$X_C - X_L = X_L$$

(note that if we put $X_L - X_C = X_L$, $X_C = 0$ which does not satisfy our problem) and $\quad X_C = 2X_L$

$$\frac{10^6}{2\pi fC} = 2 \times 2\pi fL$$

$$\therefore\ C = \frac{10^6}{2 \times 2^2\pi^2 \times 50^2 \times 0.318}$$

$$= \underline{15.9\,\mu F}$$

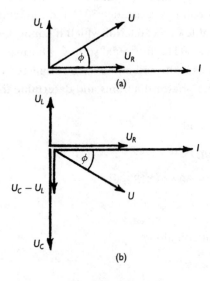

Fig. 22 Phasor diagrams: (a) for coil alone; (b) for coil and capacitor in series

The power factor has the same value for both circuit conditions; in the first case it is lagging, in the second case leading. Its value is obtained as follows:

$$\cos\phi = \frac{R}{Z} \quad \begin{array}{l}\text{(resistance of coil)}\\\text{(impedance of coil)}\end{array}$$

$$= \frac{80}{128}$$

$$= \underline{0.625}$$

EXAMPLE 4   A coil has resistance $80\,\Omega$ and inductance $0.318\,\mathrm{H}$. It is connected in series with a capacitor of $158\,\mu\mathrm{F}$ to a sinusoidal supply of $256\,\mathrm{V}$ at variable frequency. Calculate the two values of frequency for which the current in the circuit is $2\,\mathrm{A}$.

Impedance
$$Z = \frac{U}{I}$$

$$= \frac{256}{2}$$

$$= 128\,\Omega$$

$$Z^2 = R^2 + (X_L - X_C)^2$$

$$(X_L - X_C)^2 = Z^2 - R^2$$

$$= 128^2 - 80^2$$

$$= 10\,000$$

$$X_L - X_C = 100 \quad \text{(i)}$$

$$[\text{or } X_C - X_L = 100 \quad \text{(ii)}]$$

$$2\pi f L - \frac{10^6}{2\pi f C} = 100$$

$$2\pi \times 0.318 f^2 - \frac{10^6}{2\pi \times 158} = 100 f \quad \text{(after multiplying by } f\,)$$

$$2f^2 - 1000 = 100 f$$

or
$$2f^2 - 100 f - 1000 = 0$$

This is a quadratic equation which is found in the general form

$$ax^2 + bx + c = 0$$

where a, b and c are constants and its two solutions are given by

$$x = \frac{-b \pm \sqrt{b^2 - 4ac}}{2a}$$

Here $x = f$, $a = 2$, $b = -100$, $c = -1000$,

$$\therefore \quad f = \frac{100 \pm \sqrt{(-100)^2 - (4 \times 2 \times -100)}}{2 \times 2}$$

$$= \frac{100 \pm \sqrt{10\,000 + 8000}}{4}$$

$$= \frac{100 \pm \sqrt{18\,000}}{4}$$

$$= 58.6 \text{ and } -8.6$$

The negative frequency is meaningless but -8.6 satisfies the condition (ii) above.

The required frequencies are thus 58.6 and 8.6 Hz.

EXAMPLE 5 A series circuit consists of an unknown capacitor, a capacitor of 80 µF and a capacitor of 240 µF. A d.c. supply of 10 kV is applied to the circuit. The total energy stored by the circuit is 1.5 kJ.
(a) Calculate the capacitance of the unknown capacitor.
(b) Determine the voltage across each capacitor when they are fully charged.

(a) Total energy stored $(W) = \frac{1}{2}CU^2$ joules, thus
$1.5 \times 10^3 = 0.5 \times C \times 10\,000^2$

$$C_T = \frac{1500}{0.5 \times 10^{-6} \times 10\,000^2}$$

$$= 30 \, \mu\text{F}$$

Now $\quad \dfrac{1}{C_T} = \dfrac{1}{C_1} + \dfrac{1}{C_2} + \dfrac{1}{C_3}$

$$\frac{1}{30} = \frac{1}{80} + \frac{1}{240} + \frac{1}{C_3}$$

$$\frac{1}{C_3} = \frac{1}{30} - \frac{1}{80} - \frac{1}{240}$$

$$\frac{1}{C_3} = \frac{8 - 3 - 1}{240}$$

$$\frac{1}{C_3} = \frac{4}{240}$$

$$C_3 = 60\,\mu\text{F}$$

(b) Voltage across $80\,\mu\text{F}$ capacitor $= \left(\frac{3}{8} \times 10\,000\right) = 3750\,\text{V}$

Voltage across $240\,\mu\text{F}$ capacitor $= \left(\frac{1}{8} \times 10\,000\right) = 1250\,\text{V}$

Voltage across $60\,\mu\text{F}$ capacitor $= \left(\frac{4}{8} \times 10\,000\right) = 5000\,\text{V}$

EXERCISE 2

1. A resistor of $15\,\Omega$, an inductor of $0.0318\,\text{H}$ and a $150\,\mu\text{F}$ capacitor are connected in series to a $230\,\text{V}$ sinusoidal supply at $50\,\text{Hz}$.
Calculate:
(a) the impedance of the whole circuit;
(b) the current;
(c) the power factor;
(d) the voltage across each component.
    Draw the phasor diagram to scale.

2. When a coil is connected to a $240\,\text{V}$ d.c. supply a current of $2\,\text{A}$ flows. When it is connected to a $230\,\text{V}$ a.c. supply at $50\,\text{Hz}$ the current is $1\,\text{A}$. Calculate the current which flows when the same coil is connected in series with a capacitor of $30\,\mu\text{F}$ to the same a.c. supply.

3. Complete the table which refers to inductance and capacitance in series and the frequency of resonance.

Inductance (H)	0.3			1.5		1	
Capacitance (μF)	25	40			50		30
Resonant frequency (Hz)			50	100			250

4. A resistance of $24\,\Omega$, a capacitance of $160\,\mu\text{F}$ and an inductance of $0.16\,\text{H}$ are connected in series with each other. A supply at $230\,\text{V}$, $50\,\text{Hz}$ is applied to the ends of the combination.
Calculate:
(a) the current in the circuit;

(b) the potential differences across each element of the circuit;

(c) the frequency to which the supply would need to be changed so that the current would be at unity power factor, and find the current at this frequency.

5. A coil of insulated wire of resistance 8 Ω and inductance 0.03 H is connected to an a.c. supply at 240 V, 50 Hz.

   Calculate:

   (a) the current, the power, and the power factor;

   (b) the value (in microfarads) of a capacitor which, when connected to a series with the above coil, causes no change in the values of current and power taken from the supply.

6. A coil has inductance 0.223 H and resistance 50 Ω. Find the two values of capacitance which when connected in series with the coil to a 230 V 50 Hz supply will cause a current of 2.9 A to flow and calculate the corresponding power factors.

7. A circuit consists of a resistor of 35 ohms, an inductor of negligible resistance and a capacitor connected in series. A current of 2 amperes flows in the circuit when it is connected to a 50 hertz sinusoidal supply.

   A voltmeter connected in turn across the inductor and capacitor reads 200 volts and 130 volts respectively. Calculate:

   (a) the value of the inductance;

   (b) the value of the capacitance;

   (c) the supply voltage.

   Draw the phasor diagram to scale. (C & G)

8. A coil of resistance 100 ohms and inductance 0.244 henry is connected in series with a capacitor of 30 microfarad to a 100 volt sinusoidal supply of variable frequency. Calculate the two values of frequency for which the current is 0.707 ampere.

9. An inductor of 0.5 henry is connected in series with a capacitor of 72 microfarad and a resistor of unknown value.

   Calculate:

   (a) the resonant frequency of the circuit;

   (b) the value of a resistor so that at resonance the current taken from a 100 volt supply shall be 4 amperes;

   (c) the p.d. across each component at resonance.

Alternating current circuit calculations II
Parallel circuits

References: **Inductive reactance.**

Capacitive reactance.
Impedance.
Phasor resolution and combination.

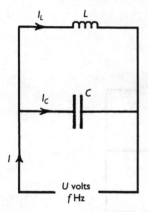

Fig. 23

Consider a circuit consisting of inductance and capacitance in parallel (Fig. 23).

L is a pure inductance (henry)
C is a pure capacitance (microfarad)

The same voltage is applied to each branch of the circuit. The current through the inductance is $I_L = U_{XL}$ where $X_L = 2\pi fL$. This current lags the voltage by 90°. The capacitor current is $I_C = U/X_C = 10^6/2\pi fC$. This current leads the voltage by 90°. The phasor diagram is usually drawn as in Fig. 24.

Usually the inductive branch possesses resistance as shown in Fig. 25. I_L lags the voltage U by some angle less than 90° depending upon the values of L and R (L and R form a series circuit).

The phasor diagram is shown in Fig. 26. ϕ_L is the phase angle of the inductive branch and $\tan\phi_L = X_L/R$ (see Volume 2).

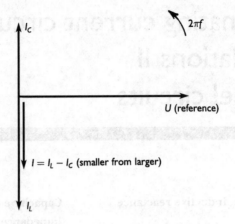

Fig. 24

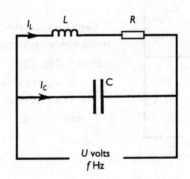

Fig. 25

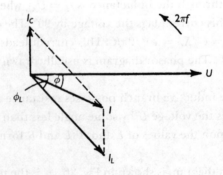

Fig. 26

The supply current is the phasor sum of I_C and I_L which is found by completing the parallelogram; ϕ is the phase angle between supply voltage and current.

EXAMPLE I A coil has resistance $25\,\Omega$ and inductive reactance $20\,\Omega$. It is connected in parallel with a capacitor of reactance $40\,\Omega$ to a $230\,V$ a.c. supply. Determine the supply current and the overall power factor.

Coil impedance

$$Z_L^2 = R^2 + X_L^2$$
$$= 25^2 + 20^2$$
$$Z_L = \sqrt{25^2 + 20^2}$$
$$= 32.02\,\Omega$$

Coil current

$$I_L = \frac{U}{Z_L}$$
$$= \frac{230}{32.02}$$
$$= 7.183\,A$$
$$= 7.2\,A$$

Coil phase angle is found from

$$\tan\phi_L = \frac{X_L}{R}$$
$$= \frac{20}{25}$$
$$= 0.8$$
$$\phi_L = 38° \, 39'$$

Capacitor current

$$I_C = \frac{U}{X_C}$$
$$= \frac{230}{40}$$
$$= 5.75\,A$$

The supply current may be determined:
(a) graphically by constructing the phasor diagram to scale as in Fig. 27.

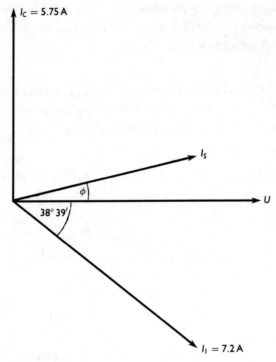

Fig. 27 Phasor diagram to scale

By measurement $I_S = 5.8$ A and $\cos \phi = 0.976$ leading or
(b) by calculation as follows:

The horizontal component of the coil current is

$$I_L \cos \phi_L = 7.2 \cos 38° 39'$$

$$= 7.2 \times 0.7810$$

$$= 5.623$$

The horizontal component of the capacitor current

$$= 0$$

The total horizontal component

$$X = 5.623 + 0$$

$$= 5.623$$

The vertical component of the capacitor current

$$= 5.75$$

The vertical component of the coil current is

$$-I_L \sin \phi_L = -7.2 \sin 38° 39'$$

$$= -7.2 \times 0.6246$$

$$= -4.497 \quad \text{(The minus sign because this component acts downwards.)}$$

The total vertical component

$$Y = 5.75 - 4.497$$

$$= 1.253$$

The resultant current

$$I_S = \sqrt{X^2 + Y^2}$$

$$= \sqrt{5.623^2 + 1.253^2}$$

$$= \underline{5.76\,\text{A}}$$

The phase angle between the supply current and the voltage is given by

$$\tan \phi = \frac{Y}{X}$$

$$= \frac{1.253}{5.623}$$

$$= 0.2228$$

$$\phi = 12° 50'$$

and the overall power factor is

$$\cos \phi = \cos 12° 50'$$

$$= \underline{0.975 \text{ leading}}$$

EXAMPLE 2   A parallel circuit consists of two branches (see Fig. 28). Branch A has inductive reactance $100\,\Omega$ and resistance $173.2\,\Omega$ in series. Branch B has capacitive reactance $173.2\,\Omega$ and resistance $100\,\Omega$ in series. The supply to the circuit is 200 V 50 Hz. Determine:

(a) the supply current and the power factor;

(b) the components of the equivalent series circuit.

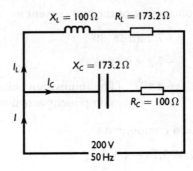

Fig. 28

For the inductive branch

$$Z_L = \sqrt{R_L^2 + X_L^2}$$

$$= \sqrt{(173.2)^2 + (100)^2}$$

$$= 200\,\Omega$$

$$I_L = \frac{U}{Z_L}$$

$$= \frac{200}{200} = 1\,\text{A}$$

The phase angle ϕ_L is found from

$$\tan\phi_L = \frac{X_L}{R} \quad \text{(see impedance triangle)}$$

$$= \frac{100}{173.2}$$

and $\qquad \phi_L = 30° \quad \text{(lag)}$

For the capacitive branch

$$Z_C = \sqrt{R_C^2 + X_C^2}$$

$$= \sqrt{(100)^2 + (173.2)^2}$$

$$= 200\,\Omega$$

$$I_C = \frac{U}{Z_C}$$

$$= \frac{200}{200} = 1\,\text{A}$$

$$\tan \phi_C = \frac{X_C}{R}$$

$$= \frac{173.2}{100}$$

$$\phi_C = 60° \quad \text{(lead)}$$

Again as in Example 1 the supply current may be determined by constructing the phasor diagram to scale or by calculation (see Fig. 29).

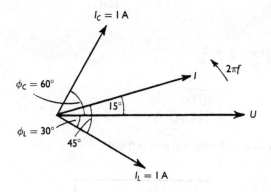

Fig. 29

It will be observed that I_L and I_C are at right angles, their resultant can thus be found by applying Pythagoras' Theorem directly:

$$I = \sqrt{I_C^2 + I_L^2}$$

$$= \sqrt{1^2 + 1^2}$$

$$= \sqrt{2} = \underline{1.414\,\text{A}} \tag{a}$$

Furthermore since $I_C = I_L$ their resultant lies at an angle of $45°$ to I_L. The phase angle between I and U is thus $45° - 30° = 15°$

(lead) and the power factor is

$$\cos\phi = \cos 15°$$
$$= \underline{0.966 \text{ lead}} \tag{a}$$

For the equivalent series circuit we require a circuit which takes a current of 1.414 A at 0.966 power factor leading from a 200 V 50 Hz supply. The circuit will consist of a capacitor and resistor in series (Fig. 30).

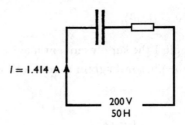

Fig. 30

Its impedance $Z = \dfrac{U}{I}$

$$= \frac{200}{1.414}$$
$$= 141.4\,\Omega$$

$\phi = 15°$, $\cos\phi = 0.966$, $\sin\phi = 0.2588$

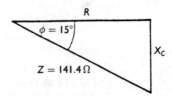

Fig. 31

From the impedance triangle (Fig. 31)

$$\cos\phi = \frac{R}{Z}$$
$$0.966 = \frac{R}{141.4}$$
$$R = 141.4 \times 0.966$$
$$= \underline{136.6\,\Omega}$$

$$\sin \phi = \frac{X_C}{Z}$$

$$0.2588 = \frac{X_C}{141.4}$$

$$X_C = 141.4 \times 0.2588$$

$$= 36.59 \, \Omega$$

$$X_C = \frac{10^6}{2\pi f C}$$

$$36.59 = \frac{10^6}{2\pi \times 50 \times C}$$

$$C = \frac{10^6}{2\pi \times 50 \times 36.59}$$

$$= 87 \, \mu F$$

The equivalent circuit thus consists of a capacitor of 87.4 μF in series with a resistor of 136.6 Ω. (b)

EXAMPLE 3 A coil has inductance 0.318 H and resistance 200 Ω. A capacitor of value 3.18 μF is connected in parallel with the coil to a sinusoidal supply of variable frequency (see Fig. 32). Calculate the frequency at which the supply current and voltage are in phase.

From the phasor diagram it is seen that for the supply current I to be in phase with the voltage U the vertical component of the soil current AB must be equal to the capacitor current.

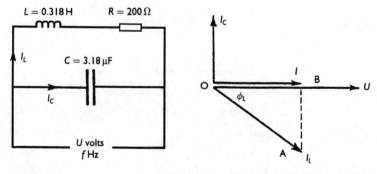

Fig. 32

Let f be the required frequency in Hz.

For the coil $X_L = 2\pi f L$

$$= 2\pi \times f \times 0.318$$

$$= 2f$$

$$Z = \sqrt{R^2 + X_L^2}$$

$$= \sqrt{200^2 + (2f)^2}$$

and $\qquad I_L = \dfrac{U}{\sqrt{200^2 + (2f)^2}}$

From the triangle OAB $\qquad \dfrac{AB}{OA} = \sin\phi_L$

$$AB = OA \sin\phi_L$$

$$= I_L \sin\phi_L$$

ϕ_L is the phase angle of the coil circuit at the frequency f. From the impedance triangle for the coil (Fig. 33)

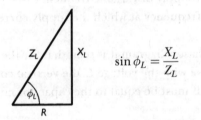

$$\sin\phi_L = \dfrac{X_L}{Z_L}$$

Fig. 33

$$\therefore \quad AB = \dfrac{U}{\sqrt{200^2 + (2f)^2}} \times \dfrac{2f}{\sqrt{200^2 + (2f)^2}}$$

$$= \dfrac{2fU}{200^2 + (2f)^2}$$

The capacitor current is $I_C = \dfrac{U}{X_C}$

and
$$X_C = \frac{10^6}{2\pi f C}$$

$$\therefore \quad I_C = \frac{U}{10^6/2\pi f C}$$

$$= U \times \frac{2\pi f C}{10^6}$$

$$= U \times \frac{2\pi \times 3.18 \times f}{10^6}$$

$$= \frac{2 \times U \times f}{10^5}$$

and
$$\frac{2fU}{200^2 + (2f)^2} = \frac{2Uf}{10^5}$$

$$\therefore \quad 200^2 + 4f^2 = 10^5$$

$$4f^2 = 100\,000 - 40\,000$$

$$f^2 = \frac{60\,000}{4}$$

$$f = \sqrt{15\,000}$$

$$= \underline{122.5\,\text{Hz}}$$

EXERCISE 3

1. A capacitor of 15 μF is connected in parallel with a coil of inductance 0.3 H and negligible resistance to a sinusoidal supply of 230 V, 50 Hz. Calculate the resultant current and state whether the phase angle is a leading or lagging one.

2. Calculate the resulting supply current and the overall power factor when a resistor of 100 Ω is connected in parallel with the circuit of question 1.

3. A coil of reactance 30 Ω and resistance 40 Ω is connected in parallel with a capacitor of reactance 200 Ω and the circuit is supplied at 200 V. Calculate the resultant current and power factor. Check the results by constructing the phasor diagram accurately to scale.

4. A parallel circuit consists of two branches. Branch A consists of a coil of resistance $100\,\Omega$ and inductance 0.552 H. Branch B consists of a capacitor of $22.5\,\mu F$ in series with a resistor of $141.4\,\Omega$. The supply to the circuit is 100 V, 50 Hz. Determine (i) by graphical construction (ii) by calculation the resultant supply current and the overall power factor.

5. Calculate the values of the components which when connected in series will form a circuit equivalent to that of question 4.

6. A resistor of 12 ohms and a capacitor of 300 microfarads are connected in series. An inductive coil of inductance 0.5 henry and resistance 8 ohms is connected in parallel with the above.
   A single-phase supply at 230 V, 50 Hz is connected across the ends of the combination.

   Determine graphically or by calculation,

   (a) the current and its power factor in each of the parallel circuits;

   (b) the total current from the supply and its power factor.

7. A coil has resistance $150\,\Omega$ and inductance 0.478 H. Calculate the value of a capacitor which when connected in parallel with this coil to a 50 Hz supply will cause the resultant supply current to be in phase with the voltage.

8. The two branches of a parallel circuit consist respectively of a coil of inductance 0.636 H and resistance $300\,\Omega$ and a capacitor of $3\,\mu F$. Determine the frequency at which the supply voltage and current will be in phase.

9. An inductive coil of resistance $50\,\Omega$ takes a current of 1 A when connected in series with a capacitor of $31.8\,\mu F$ to a 230 V 50 Hz supply. Calculate the resultant supply current when the capacitor is connected in parallel with the coil to the same supply.

10. A coil of resistance $50\,\Omega$ and inductance 0.276 H is connected to a sinusoidal supply at 240 V, 50 Hz. Calculate or determine graphically the value of a capacitor in microfarad which when connected in parallel with the coil will cause no change in the value of the supply current or overall power factor.

Three-phase circuit calculations I
Star connections

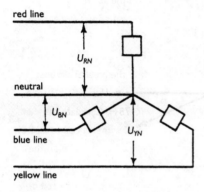

Fig. 34

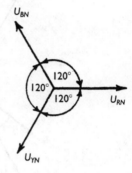

Fig. 35

Figure 34 shows three loads connected in the star formation to a three-phase four-wire supply system. Figure 35 shows the phasor diagram; the red line to neutral voltage U_{RN} is taken as reference and the phase sequence is Red, Yellow, Blue so that the other line to neutral voltages (or *phase* voltages) lie as shown.

If $U_{RN} = U_{YN} = U_{BN}$ and they are equally spaced the system of *voltages* is balanced.

Let U_L be the voltage between any pair of lines (the *line* voltage) and

$$U_P = U_{RN} = U_{YN} = U_{BN} \quad \text{(the \textit{phase} voltage)}$$

Then $U_L = \sqrt{3}U_P$

and $I_L = I_P$

where I_L is the current in any line and I_P is the current in any load or phase. The power per phase is $P = U_P I_P \cos \phi$ and the total power is the sum of the amounts of power in each phase.

If the currents are equal and the phase angles are the same, as in Fig. 36, the load on the system is balanced, the current in the neutral is zero and the total power is

$$P = \sqrt{3} U_L I_L \cos \phi$$

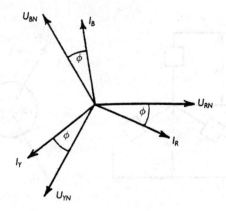

Fig. 36

EXAMPLE I The following loads are connected to a 400 V, 50 Hz three-phase four-wire system (Fig. 37):

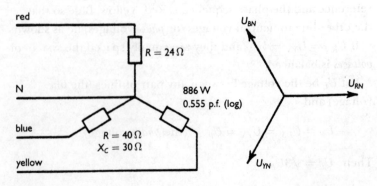

Fig. 37

between red line and neutral, a non-inductive resistor of 24 Ω;
between yellow line and neutral, 886 W at 0.555 p.f. lagging;
between blue line and neutral, a capacitor of reactance 30 Ω in
series with a resistor of 40 Ω.

The phase sequence is red, yellow, blue. Calculate:

(a) the current in each line;
(b) the total power;
(c) the current in the neutral.

Line current = phase current

$$= \frac{\text{line to neutral voltage}}{\text{impedance between line and neutral}}$$

Line to neutral voltage $= U_P = \dfrac{1}{\sqrt{3}} \times U_L$

$$= \frac{400}{\sqrt{3}}$$

$$= 230 \, \text{V}$$

Current in red line $\quad I_R = \dfrac{230}{24}$

$$= \underline{9.58 \, \text{A}} \qquad\qquad\qquad \text{(a)}$$

The power in the yellow phase circuit

$$P_Y = U_{YN} I_Y \cos \phi_Y$$

$$886 = 230 \times I_Y \times 0.555$$

Current in yellow line

$$I_Y = \frac{886}{230 \times 0.555}$$

$$= \underline{6.94 \, \text{A}} \qquad\qquad\qquad \text{(a)}$$

The phase angle between this current and the yellow to neutral
voltage is given by

$$\cos \phi_Y = 0.555$$

$$\phi_Y = 56° \, 17' \quad \text{(lag)}$$

Current in the blue line

$$I_B = \frac{230}{\sqrt{R^2 + X_C^2}}$$

$$= \frac{230}{\sqrt{40^2 + 30^2}}$$

$$= \underline{4.6\,A} \qquad\qquad\text{(a)}$$

The phase angle between this current and the blue to neutral voltage is given by

$$\tan\phi_B = \frac{X_C}{R} = \frac{30}{40}$$

$$= 0.75$$

$$\phi_B = 36°\,52' \quad \text{(lead)}$$

Power in the red phase $\quad P_R = 230 \times 10$

$$= 2300\,\text{W}$$

Power in the blue phase $P_B = 230 \times 4.8 \times \cos 36°\,52'$

$$= 230 \times 4.8 \times 0.8$$

$$= 883.2\,\text{W}$$

Total power $\qquad\qquad\qquad P = 2300 + 886 + 883.2$

$$= \underline{4069.2\,\text{W}} \qquad\qquad\text{(b)}$$

To determine the neutral current graphically draw the phasor diagram of currents accurately to scale as in Fig. 38.

Note

$$I_N = I_R + I_Y + I_B$$

means that I_N is the *phasor* sum of I_R, I_Y and I_B.

The phasors representing I_Y and I_B are first combined by completing the parallelogram. Their resultant is combined with phasor I_R to give the neutral current I_N which is determined by measurement.

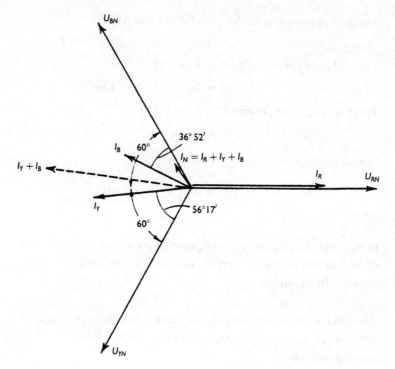

Fig. 38

Alternatively the neutral current may be calculated as follows:

Horizontal component of I_R = 9.583

Horizontal component of I_Y $= -6.94 \cos{(60° - 56° 17')}$

$\qquad\qquad\qquad\qquad\qquad = -6.94 \cos 3° 43'$

$\qquad\qquad\qquad\qquad\qquad = -6.94 \times 0.9979 = -6.925$

Horizontal component of I_B $= -4.6 \cos{(60° - 36° 52')}$

$\qquad\qquad\qquad\qquad\qquad = -4.6 \cos 23° 8'$

$\qquad\qquad\qquad\qquad\qquad = -4.6 \times 0.9196 = 4.23$

Total horizontal component $X = 10 - 6.925 - 4.23$

$\qquad\qquad\qquad\qquad\qquad = -1.155$

Vertical component of I_R $= 0$

Vertical component of I_Y $= -6.925 \sin 3° 43'$

$$= -6.925 \times 0.0648 = -0.449$$

Vertical component of I_B $= 4.6 \sin 23° 8'$

$$= 4.6 \times 0.3928 = 1.8$$

Total vertical component $Y = 1.8 - 0.449$

$$= 1.351$$

The neutral current I_N $= \sqrt{X^2 + Y^2}$

$$= \sqrt{(1.155)^2 + (1.351)^2}$$

$$= \underline{1.777\,\text{A}} \qquad\qquad\qquad (c)$$

Remember that horizontal components are negative if they lie to the left of the origin 0. Vertical components are negative if they lie beneath the origin 0.

EXAMPLE 2   The following three-phase loads are supplied at 400 V, 50 Hz to a group of three buildings by the same distribution cable:

   50 kW at unity power factor;
   80 kVA at 0.6 power factor leading;
   40 kVA at 0.7 power factor lagging.

Find by calculation or graphically:

(a) the total load in kilovolt amperes (kVA);
(b) the total kilowatts;
(c) the line current in the cable;
(d) the combined power factor.   [C & G (modified)]

(a) 50 kW at unity power factor

   active component $= 50\,\text{kW}$

   reactive component $= 0$

   80 kVA at 0.6 lagging power factor

   active component $= 80 \cos \phi_2$

   $$= 80 \times 0.6$$

   $$= 48\,\text{kW}$$

reactive component $= 80 \sin \phi_2$

$$= 80 \times 0.8$$

$$= 64 \,\mathrm{kVA}, \quad \mathrm{lag}$$

40 kVA at 0.7 leading power factor

active component $\;= 40 \cos \phi_1$

$$= 40 \times 0.7$$

$$= 28 \,\mathrm{kW}$$

reactive component $= 40 \sin \phi_1$

$$= 40 \times 0.71$$

$$= 28.4 \,\mathrm{kVA}, \quad \mathrm{lead}$$

Tabulating the values:

Load	Active components (kW)	Reactive components (kVAr)
50 kW	50	0
80 kVA	48	64
40 kVA	28	−28.4
Resultant components	126	35.6

Then total kVA $= \sqrt{\mathrm{kW}^2 + \mathrm{kVA}^2}$

$$= \sqrt{126^2 + 35.6^2}$$

$$= 131 \,\mathrm{kVA}$$

(b) Load in kilowatts from above 126 kW

(c) Line current $= \dfrac{\mathrm{kVA} \times 10^3}{\sqrt{3} \times 400}$

$$= \dfrac{131 \times 10^3}{\sqrt{3} \times 400}$$

$$= 189 \,\mathrm{A}$$

(d) Combined power factor $= \dfrac{\text{kW}}{\text{kVA}}$

$$= \dfrac{126}{131}$$

$$= 0.962 \quad \text{lag}$$

EXERCISE 4

1. Three equal coils of inductive reactance $30\,\Omega$ and resistance $40\,\Omega$ are connected in star to a three-phase supply of line voltage 400 V. Calculate the line current and the total power.

2. The load connected between each line and the neutral of a 400 V, 50 Hz three-phase circuit consists of a capacitor of $31.8\,\mu\text{F}$ in series with a resistor of $100\,\Omega$. Calculate the line current and the total power.

3. The load connected between each line and the neutral of a 400 V three-phase supply consists of:

    between red line and neutral, a non-inductive resistance of $25\,\Omega$;

    between yellow line and neutral, an inductive reactance $12\,\Omega$ in series with resistance $5\,\Omega$;

    between blue line and neutral, a capacitive reactance $17.3\,\Omega$ in series with resistance $10\,\Omega$.

    Calculate the current in each line and the total power.

4. A star-connected resistance bank, each resistor of $30\,\Omega$ is connected to a 400 V three-phase supply. Connected to the same supply is a star-connected capacitor bank, each capacitor having reactance $40\,\Omega$. Calculate the resultant current in each line and the total power.

5. The load connected between each line and neutral of a 400 V three-phase supply system consists of:

    between red line and neutral, 40 ohms resistance;

    between yellow line and neutral, 20 ohms resistance;

    between blue line and neutral, 60 ohms resistance.

    Calculate the current in the neutral and check the result by means of an accurately constructed phasor diagram. Calculate also the total power supplied. (C & G)

6. A 400 V, three-phase, four-wire system supplies power to three non-inductive loads. The loads are 25 kW between red and neutral, 30 kW between yellow and neutral, and 12 kW between blue and neutral.

Calculate (a) the current in each line wire, and (b) the current in the neutral conductor. (C & G)

7. The load connected between each line and neutral of a 400 V three-phase four-wire system is as follows:

red line	12 kW 0.866 p.f. lagging
yellow line	10 kW unity p.f.
blue line	8 kW 0.707 p.f. leading

Determine graphically or by calculation the current in the neutral.

8. The load connected between each line and the neutral of a 400 V supply system consists of:

between red line and neutral, 100 Ω non-inductive resistance;

between yellow line and neutral, 100 Ω inductive reactance;

between blue line and neutral, 100 Ω capacitive reactance.

The phase sequence is RYB. Calculate the current in the neutral.

9. Use a graphical construction to determine the neutral current of question 8 when the phase sequence is reversed.

10. The circuit shown in Fig. 39 is connected to a 400 V 50 Hz supply with phase sequence RYB. Calculate or determine graphically the current in the neutral.

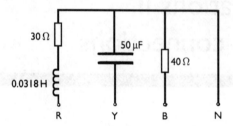

30 Ω

50 μF

40 Ω

0.0318 H

R   Y   B   N

Fig. 39

11. For the parallel circuit shown in Fig. 40

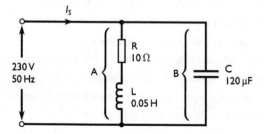

I_S

230 V
50 Hz

A

R
10 Ω

L
0.05 H

B

C
120 μF

Fig. 40

(a) calculate for branch A, the:
  (i)   impedance
  (ii) current
  (iii) power factor
  (iv) phase angle;
(b) calculate for branch B, the:
  (i)   reactance
  (ii) current;
(c) by calculation or by a suitably scaled phasor diagram find the:
  (i)   total current
  (ii) overall power factor.

Three-phase circuit calculations II
Delta connections

Figure 41 shows three loads connected in the delta or mesh formation to a three-phase supply system. Figure 42 shows the phasor diagrams of the line voltages with the red to yellow voltage taken as reference.

The voltage applied to any load is the line voltage U_L and the line current is the phasor difference between the currents in the two loads connected to that line. In particular if the load currents are all equal and make equal phase angles with their respective voltages the system is balanced and

$$I_L = \sqrt{3}I_P$$

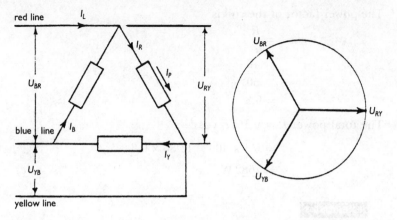

red line

I_L

I_R

I_P

U_{BR}

U_{RY}

blue line

I_B

U_{YB}

I_Y

yellow line

U_{BR}

U_{RY}

U_{YB}

Fig. 41   **Fig. 42**

The total power under these conditions is

$$P = \sqrt{3}\,U_L I_L \cos\phi$$

EXAMPLE   Three coils each of resistance $40\,\Omega$ and inductive reactance $30\,\Omega$ are connected in delta to a $400\,V$ three-phase system.

Calculate:

(a) the current in each coil,

(b) the line current,

(c) the total power.

The circuit diagram is as in Fig. 41.

Impedance of each coil $Z = \sqrt{R^2 + X_L^2}$

$$= \sqrt{40^2 + 30^2}$$

$$= 50\,\Omega$$

Current in each coil $\qquad = \dfrac{U}{Z}$

$$= \dfrac{400}{50} = \underline{8\,A} \qquad\qquad \text{(a)}$$

Line current $\qquad\qquad I_L = \sqrt{3} \times 8$

$$= \underline{13.86\,A} \qquad\qquad \text{(b)}$$

The power factor of the coil is

$$\cos \phi = \frac{R}{Z}$$
$$= \frac{40}{50}$$
$$= 0.8$$

The total power $P = \sqrt{3} U_L I_L \cos \phi$

$$= \sqrt{3} \times 400 \times 13.86 \times 0.8$$
$$= \underline{7682\,\text{W}} \qquad\qquad\qquad (c)$$

EXERCISE 5

1. Three resistors each of 30 Ω are connected
   (a) in star,
   (b) in delta,
   to a 400 V three-phase system. Calculate the current in each resistor, the line current and the total power for each connection.
2. Each branch of a mesh-connected load consists of resistance 20 Ω in series with inductive reactance 30 Ω. The line voltage is 400 V. Calculate the line current and the total power.
3. Three coils each with resistance 45 Ω and with inductance 0.2 H are connected to a 400 V, three-phase supply at 50 Hz
   (a) in mesh, (b) in star.
   Calculate for each method of connection
   (i) the current in each coil, and
   (ii) the total power in the circuit. (C & G)
4. A three-phase load consists of three similar inductive coils, each of resistance 50 Ω and inductance 0.3 H. The supply is 415 V, 50 Hz. Calculate:
   (a) the line current;
   (b) the power factor;
   (c) the total power;
   when the load is (i) star-connected, (ii) delta-connected. (C & G)
5. Three equal resistors are required to absorb a total of 24 kW from a 400 V three-phase system. Calculate the value of each resistor when they are connected
   (a) in star, (b) in mesh.

6. To improve the power factor of a certain installation requires a total of 48 kVAr equally distributed over the three phases of a 415 V 50 Hz system. Calculate the value of the capacitors required (in microfarads) when the capacitors are connected

    (a) in star, **(b)** in delta.

7. The following loads are connected to a three-phase three-wire 400 V 50 Hz supply system:

    between red and yellow lines a non-inductive resistance 60 Ω;

    between yellow and blue lines, a coil of inductive reactance 30 Ω and resistance 40 Ω;

    between blue and red lines, a capacitor of 100 μF.

    Calculate the current through each load and the total power.

8. A 400 V three-phase star-connected alternator supplies a delta-connected induction motor of full load efficiency 87% and power factor 0.8 which delivers 14 920 W. Calculate:

    (a) the current in each motor winding;

    (b) the current in each alternator winding;

    (c) the power to be developed by the engine driving the alternator assuming that the efficiency of the alternator is 82%.

Three-phase power and power factor improvement

References: **Watts**.
              Voltamperes.
              Reactive voltamperes.
              Tariffs.

EXAMPLE I A works load consists of

(i)   9 kW of lighting at unity p.f.;

(ii)   a motor taking 12 kVA at 0.75 p.f. lagging;

(iii) a number of small motors taking 15 kW at 0.6 p.f. lagging.

The loads are balanced over the three phases of a 400 V supply system. Determine:

(a) the total kW;

(b) the total kVAr;

(c) the overall kVA;

(d) the overall power factor;

(e) the line current.

The power triangle for a lagging power factor load is as shown in Volume 2 (see Fig. 43).

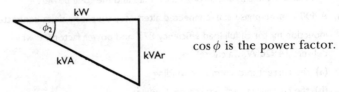

$\cos \phi$ is the power factor.

Fig. 43

For load (ii) 12 kVA at 0.75 p.f. lagging:

$$\cos \phi = 0.75, \ \phi = 41° 24', \ \sin \phi = 0.6613$$

true power $= kVA \times$ power factor

$$= 12 \times 0.75$$

$$= 9 \, kW$$

Reactive kVAr $= kVA \sin \phi$

$$= 12 \times 0.6613$$

$$= 7.9356$$

$$= 7.936$$

For load (iii) $\cos \phi = 0.6$, $\phi = 53° 8'$, $\tan \phi = 1.3335$

$$\frac{kVAr}{kW} = \tan \phi$$

$$kVAr = kW \tan \phi$$

$$= 15 \times 1.3335$$

$$= 20$$

These results may then be tabulated:

Load	kW	kVAr	
i	9	0	
ii	9	7.936	
iii	15	20	
Total	33	27.936	(a) (b)

Note Only kW and kVAr may be added directly.

The combined load is then represented by a power triangle drawn to scale if required (Fig. 44).

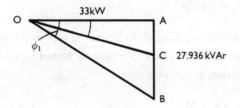

Fig. 44

Overall kVA $= \sqrt{33^2 + 27.936^2}$ (or by measurement)

$$= \underline{43.23\,\text{kVA}} \qquad \text{(c)}$$

The overall power factor

$$\cos \phi_1 = \frac{\text{kW}}{\text{kVA}}$$

$$= \frac{33}{43.23}$$

$$= \underline{0.763} \quad \text{lag} \qquad \text{(d)}$$

Since VA $= \sqrt{3}U_L I_L$ in a three-phase system

The line current $I_L = \dfrac{\text{VA}}{\sqrt{3} \times U_L}$

$$= \frac{43.23 \times 1000}{\sqrt{3} \times 400}$$

$$= \underline{62.4\,\text{A}} \qquad \text{(e)}$$

Note The calculations may also be performed on a single-phase basis if desired but since the loads are balanced this is not really necessary.

EXAMPLE 2 Calculate:

(a) the total kVAr to be supplied by a capacitor bank in order to improve the overall power factor of the system of Example 1 to 0.9 p.f. lagging;

(b) the value of capacitance required assuming that the capacitors are connected (i) in star, (ii) in delta.

Using the accurately drawn power triangle of the previous example and inserting the additional line OC set off from OA at an angle ϕ_2 given by

$$\cos\phi_2 = 0.9$$

$$\phi_2 = 25° \, 50'$$

BC represents the leading kVAr required to bring about the desired improvement. The kVAr may be found by measuring BC or by calculation as follows:

$$BC = AB - AC$$

as previously determined

$$AB = 27.936$$

Similarly $\quad AC = 33 \tan 25° \, 50'$

$$= 33 \times 0.4841$$

$$= 15.97$$

$$BC = 27.936 - 15.97$$

$$= \underline{11.966} \qquad \text{(a)}$$

This is the total kVAr required.

$$\text{kVAr required per phase} = \frac{11.966}{3}$$

$$= 3.989 \text{ or } 3989 \, \text{VAr}$$

Since in a capacitive circuit

$$I = \frac{U}{X_C}$$

multiply both sides by U so that

$$UI = \frac{U^2}{X_C} \quad \text{(volt amperes reactive)}$$

Thus $\quad \dfrac{U^2}{X_C} = 3989$

For the star connection $U = \dfrac{1}{\sqrt{3}} \times 400$

$$= 230\,\text{V}$$

and $\quad \dfrac{230^2}{X_C} = 3989$

$$X_C = \frac{230^2}{3989}$$

$$= 13.26\,\Omega$$

$$\frac{10^6}{2\pi f C} = 13.26$$

$$C = \frac{10^6}{2\pi \times 50 \times 13.26}$$

$$= \underline{240\,\mu\text{F}}$$

so that three capacitors each of $240\,\mu\text{F}$ connected in star would be required. (b) (i).

For the delta connection $U = 400\,\text{V}$

and $\quad \dfrac{400^2}{X_C} = 3989$

$$X_C = \frac{400^2}{3989}$$

$$= 40.1\,\Omega$$

$$C = \frac{10^6}{2\pi \times 50 \times 40.1}$$

$$= \underline{79.4\,\mu\text{F}}$$

so that three capacitors each of $79.4\,\mu\text{F}$ connected in delta would be required. (b) (ii).

EXAMPLE 3 The power taken by a three-phase 400 V 50 Hz induction motor is 85 kW at a power factor of 0.72 lagging. A bank of capacitors is to be connected in delta across the motor terminals to improve the overall power factor to 0.9 lagging. Determine the required capacitance per phase.

kVA required = kW $(\tan\phi_1 - \tan\phi_2)$

Now

$$\cos\phi_1 = 0.72 = 43.9° \quad \text{and} \quad \tan\phi_1 = 0.964$$

and

$$\cos\phi_2 = 0.9 = 25.8° \quad \text{and} \quad \tan\phi_2 = 0.484$$

Then kVAr $= 85\,(0.964 - 0.484)$

$$= 40.8 \text{ kVAr}$$

Line current of capacitor bank $= \dfrac{\text{kVAr} \times 10^3}{\sqrt{3} \times U_L}$

$$= \dfrac{40\,800}{\sqrt{3} \times 400}$$

$$= 58.9 \text{ A}$$

As the capacitors are in delta the capacitor current

$$= \dfrac{56.9}{\sqrt{3}}$$

$$= 34 \text{ A}$$

Reactance of capacitor (per phase)

$$X_C = \dfrac{400}{34}$$

$$= 11.76\,\Omega$$

Capacitance $C = \dfrac{10^6}{2\pi f \times X_C}$

$$= \dfrac{10^6}{314 \times 11.76}$$

$$= 270\,\mu\text{F}$$

EXAMPLE 4   Two alternative tariffs are offered by a supplier for domestic consumers: either a two-part tariff consisting of a standing charge of 23.5 p per day plus a unit charge of 7.3 p per kWh, or a block rate comprising a fixed charge of 29.1 p per day plus a unit charge of 5.2 p per kWh for daytime use and 3.9 p per kWh for night use. On the assumption that 40% of the units used will be daytime and the remainder nighttime use, and assuming 5% VAT is chargeable, find:

(a) the total cost of each tariff for 90 days for 1000 units consumed;

(b) the average overall cost per unit on each tariff.

(a) Cost of 1000 units on two-part tariff:

Standing charge $= 90 \times 23.5\,\text{p}\ = £21.15$

Unit charge $\quad = 1000 \times 7.3\,\text{p} = \underline{£73.00}$

$\qquad\qquad\qquad$ Total $\quad£94.15 \times 1.05$

$\qquad\qquad$ inc. VAT $= £98.86$

Cost of 1000 units on block tariff:

Standing charge $= 90 \times 29.1\,\text{p} = £26.19$

Daytime units $\quad = 400 \times 5.2\,\text{p} = £20.80$

Nighttime units $\ = 600 \times 3.9\,\text{p} = \underline{£23.40}$

$\qquad\qquad\qquad$ Total $\quad£70.39 \times 1.05$

$\qquad\qquad$ inc. VAT $= £73.91$

(b) Overall cost per unit:

On two-part tariff $\dfrac{9886}{1000} = 9.886\,\text{p}$

On block tariff $\quad\dfrac{7391}{1000} = 7.391\,\text{p}$

EXAMPLE 5   An industrial consumer with a constant maximum demand throughout the year is offered the following tariff: £17.00 per kW of maximum demand per annum, plus 5.5 p per unit consumed. The tariff also includes a 'power factor clause' to the effect that 'the amount payable for each kW of maximum demand shall be increased by 1 per cent for each 0.01 by which the average lagging power factor is less than 0.9'.

If the consumer's maximum demand is 280 kW with an annual consumption of 500 000 kWh, and his average power factor is 0.75 lagging, calculate:

(a) the total annual cost including 5% VAT;

(b) the overall cost per unit.

(a) Increase in kW of maximum demand charge $= 0.9 - 0.75$

$$= 0.15$$

Percentage increase of maximum demand charge $= \dfrac{0.15 \times 1\%}{0.1}$

$$= 15\%$$

So charge per kW of maximum demand $= £17.00 + \dfrac{15 \times £17.00}{100}$

$$= £17.00 + £2.55$$

$$= £19.55$$

Annual charge for stated maximum demand $= £19.55 \times 280$

$$= £5474$$

Annual kWh charge $= 500\,000 \times 5.5 = £27\,500$

Total annual cost $= £5474 + £27\,500$

$$= £32\,974 \times 1.05$$

$$= £34\,622.70$$

(b) Cost per unit $= \dfrac{34\,622.7 \times 100}{500\,000}$

$$= 6.925\,\text{p per kWh}$$

EXAMPLE 6 Calculate the cost per year of the energy supplied to a factory which is loaded daily as follows: 250 kVA for 2 hours; 180 kVA for 8 hours, and 75 kVA for 6 hours per day. The charge for the energy is made on the basis of £26 per kVA of maximum demand plus 5.8 p per unit. VAT is chargeable at 5%.

Maximum demand = 250 kVA

Maximum demand charge = 250 × £26 = £6500.

Assuming a 5-day week and 50-week year the total number of units consumed in one year is

$$250 \times 2 \times 5 \times 50 = 125\,000$$

$$+180 \times 8 \times 5 \times 50 = 360\,000$$

$$+75 \times 6 \times 5 \times 50 = \underline{112\,500}$$

$$\text{Total} \quad 597\,500$$

$$\text{Annual cost of units} = 597\,500 \times 5.8\,\text{p}$$

$$= 3465\,500\,\text{p}$$

$$= £34\,655.00$$

Basic annual cost $\quad = £6500 + £34\,655 = £41\,155$

Total annual cost $\quad = £41\,155 \times 1.05$

Including VAT $\quad = \underline{£43\,212.75}$

EXAMPLE 7 Calculate the annual cost of the energy supplied to the installation of Example 6 if additionally there is a power factor penalty clause which allows for the cost per kVA of maximum demand to be increased by £1.90 for every 0.1 by which the power factor falls below 0.85 and the average power factor is 0.7.

Determine also the overall cost per unit under these conditions.

Additional cost per kVA of maximum demand:

$$= \frac{(0.85 - 0.7)}{0.1} \times £1.90$$

$$= \frac{0.15}{0.1} \times £1.90$$

$$= £2.85$$

New maximum demand charge = 250 × £28.85 = £7212.50

Total annual cost $= £7212.50 + £34\,655 = £41\,867.50 \times 1.05$

$$= £43\,960.87$$

Overall cost per unit $= \dfrac{£43\,960.87}{597\,500}$ p

$$= \underline{7.36\,\text{p}}$$

1. An installation supplies the following loads:
   (a) 10 kW at unity power factor;
   (b) 15 kVA at 0.8 p.f. lagging;
   (c) 4 kVAr leading.
   Calculate the total kW, kVAr, the overall kVA and power factor.
2. The following loads are balanced over the three phases of a 400 V supply system:
   (a) 20 kVA at 0.8 p.f. lagging;
   (b) 25 kVA at 0.6 p.f. lagging;
   (c) 30 kW at unity p.f.
   Calculate the overall power factor and the line current.
3. (a) Calculate the line current taken by a 400 V, three-phase motor working at full load output of 14 920 W when its efficiency is 85% and power factor 0.7.
   (b) Determine the line current and resultant power factor when a capacitor bank of 8 kVAr is connected in parallel with the motor.
4. A 400 V three-phase system supplies the following balanced loads:
   (a) 8 kW of lighting at unity power factor;
   (b) a motor of full load efficiency 80%, power output 7460 W and power factor 0.75 lagging;
   (c) a number of small motors of output totalling 8952 W, efficiency 70% and power factor 0.7 lagging.
   Determine the total load in kW, kVAr, and kVA, the overall power factor and the line current.
5. Determine the:
   (a) total leading kVAr required to improve the power factor of question 4 to 0.9 lagging;

(b) values of capacitors (in microfarads) required to supply the kVAr if the capacitors are connected (i) in star, (ii) in delta. (Supply frequency 50 Hz.)

6. The load in a small works consists of 20 kW at unity power factor, and a number of small single-phase motors of output totalling 44 760 W, working at 86% efficiency and at a power factor of 0.7 lagging.

Find, graphically or by calculation the:

(a) combined load in kVA;

(b) overall power factor;

(c) load in kW;

(d) total current taken from a 230 V single-phase supply.　　　(C & G)

7. A consumer is supplied with electric power at 400 V, three-phase, four-wire. The total load consists of:

(a) 50 kW for heating and lighting at unity power factor;

(b) 90 kVA of induction motors at 0.7 power factor lagging;

(c) 30 kVA to a rotary converter at 0.6 power factor leading;

each of these loads being balanced across the three phases.

Find the value of the line current, and the power factor of the combined load.　　　(C & G)

8. Explain the meaning of power factor, and use a phasor diagram to illustrate power factor improvement.

The power taken by a 415 V, 50 Hz, three-phase induction motor is 60 kW at 0.75 power factor lagging. A bank of capacitors is connected in delta across the supply lines to improve the overall power factor.

Calculate the capacitance per phase and the total capacitance required to raise the power factor to 0.9 lagging.　　　(C & G)

9. (a) The power supply to a 415 V, 50 Hz, three-phase induction motor is 50 kW at 0.72 power factor lagging. A bank of capacitors is connected in mesh across the supply lines to improve the overall power factor.

Calculate the capacitance per phase in order to raise the power factor to 0.9 lagging.

(b) Describe briefly a different method of power factor improvement which could be used in a large works to improve the overall power factor.　　　(C & G)

10. Complete the table which refers to the loads supplied by a three-phase 400 V system. Each load is balanced over the three phases.

Load	kVA	kW	kVAr	power factor	line current
a	15	12			
b		12		1.0	
c		0	8		
d				0.8 lag	20

Overall values.

11. Explain with connection diagram how two single-phase wattmeters may be used to measure the power supply to a three-phase load.

   The steady readings on two such wattmeters are 14 kW and 35 kW. Calculate the:

   (a) power in kW taken by the load;

   (b) power factor;

   (c) load in kVA, and

   (d) line current, if the supply is at 400 V, three-phase.

   What assumptions would be made if *one* single-phase wattmeter only were to be used? Show, with a diagram, how the instrument would be connected.

12. The cost of electrical power to a consumer is £25.60 per annum per kVA of maximum demand, plus 5.5 p per unit. VAT is chargeable at 5%.

   A consumer's maximum demand is 450 kW at 0.72 power factor lagging, and his annual consumption is 720 000 kWh.

   (a) Calculate the overall cost per unit.

   (b) Describe, giving reasons, one method by which the consumer could reduce the cost of his power whilst taking the same number of units.

13. **(a)** A consumer has a lighting load of 3.6 kW, and is to install some electric heaters. The following alternative tariffs are available (VAT is chargeable at 5%):

   Two part: £10.05 per quarter, plus 6.9 p per unit for all units consumed.

   Flat rate: 11.29 p per unit for lighting and 6.46 p for heating.

   Assuming that all the apparatus will have an average use of 4 hours daily throughout the year, calculate the kW rating of the proposed electric heaters, so that the annual cost of electricity on each tariff shall be equal.

   (b) A large industrial consumer pays for energy under the following tariff:

£27.60 per annum per kVA of maximum demand plus 6.3 p per unit. VAT is chargeable at 5%.

Explain briefly why this tariff is used.

14. (a) Justify the general use of two-part tariffs in electricity supply. Give details of one form of domestic two-part tariff.

(b) A power consumer with a constant maximum demand throughout the year is offered the following tariff: £27.05 per kW of maximum demand per annum plus 4.6 p per unit. The tariff also includes a power factor clause to the effect that 'the amount payable for each kW of maximum demand shall be increased by 1% for each 0.01 by which the average lagging power factor is less than 0.9'. The annual maximum demand is 300 kW, the average power factor is 0.7 lagging, and the annual consumption is 600 000 units. VAT is charged at 5%. Calculate:

(i) the annual cost and cost per unit, when the power factor remains at 0.7 lagging;

(ii) the annual cost and cost per unit, if the average power factor were improved to 0.9 lagging.

Voltage drop calculations in three-phase circuits

References: **Resistance of conductors.**
British Standard 7671:1992.
Requirements for Electrical Installations.

Consider a balanced load connected to a three-phase system as shown in Fig. 45, so that the line current is I ampere. Each core of the cable is assumed to have resistance r ohm and negligible

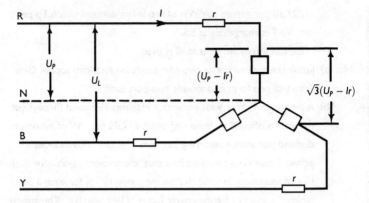

Fig. 45

reactance. The cable resistance is shown in series with each leg of the load.

Let phase and line voltages at the supply end be U_P and U_L respectively. The voltage drop in each line is Ir and, if the cable reactance and the load power factor are ignored, the phase voltage at the load end is $U_P - Ir$.

The line voltage at the supply end is $U_L = \sqrt{3}U_P$ and the line voltage at the load end is

$$\sqrt{3} \times \text{phase voltage} = \sqrt{3}(U_P - Ir)$$
$$= \sqrt{3}U_P - \sqrt{3}Ir$$
$$= U_L - \sqrt{3}Ir$$

Thus the voltage drop referred to the line voltage or total voltage drop is $\sqrt{3}Ir$.

EXAMPLE I A balanced load of 35 A is supplied over a distance of 250 m through a cable each core of which has resistance 1.351 Ω per 1000 m. The line voltage at the supply end is 400 V. Calculate the line voltage at the load end and determine the percentage drop in the line voltage.

$$\text{Resistance per core} = \frac{250}{1000} \times 1.351$$
$$= 0.3378\,\Omega$$

Voltage drop per core $= 35 \times 0.3378$

$\qquad\qquad\qquad\qquad = 11.823\,\text{V}$

Equivalent drop in the line voltage (the *total* drop)

$\qquad\qquad\qquad\qquad = \sqrt{3} \times 11.823$

$\qquad\qquad\qquad\qquad = 20.48\,\text{V}$

Line voltage at load end $\quad = 400 - 20.48$

$\qquad\qquad\qquad\qquad = \underline{379.52\,\text{V}}$

Percentage drop in line voltage $= \dfrac{20.48}{400} \times 100$

$\qquad\qquad\qquad\qquad = \underline{5.12\%}$

EXAMPLE 2 A three-phase 10 kW motor operates on full load with efficiency 80% and power factor 0.75. It is supplied from a switchboard through a cable each core of which has resistance 0.2 Ω. Calculate the voltage necessary at the supply end in order that the voltage at the load end terminals shall be 400 V.

The full load current of the motor is

$$I = \frac{10 \times 1000}{\sqrt{3} \times 400 \times 0.75} \times \frac{100}{80}$$

$$= \underline{24.06\,\text{A}}$$

The voltage drop per core of the cable

$$= 24.06 \times 0.2$$

$$= 4.8\,\text{V}$$

The equivalent reduction in the line voltage

$$= \sqrt{3} \times 4.8$$

$$= 8.31\,\text{V}$$

The required line voltage at the switchboard

$$= 400 + 8.31$$

$$= \underline{408.3\,\text{V}}$$

This is not a rigid treatment of the problem but the method gives a result sufficiently accurate for most practical purposes.

EXAMPLE 3 The estimated load in a factory extension is 50 kW balanced at 0.8 p.f. The supply point is 120 metres away and the supply voltage is 400 V. Calculate the cross-sectional area of the cable in order that the total voltage drop shall not exceed 2.5% of the supply voltage.

Take the resistivity of copper as $1.78 \times 10^{-8}\,\Omega\text{m}$.

The line current

$$I = \frac{50 \times 1000}{\sqrt{3} \times 400 \times 0.8}$$

$$= 90.21\,\text{A}$$

Allowable reduction in line voltage

$$= 2.5\% \times 400$$

$$= 10\,\text{V}$$

Equivalent reduction in phase voltage

$$= \frac{10}{\sqrt{3}}$$

$$= 5.77\,\text{V}$$

Resistance per core of the cable

$$= \frac{5.77}{90.21}$$

$$= 0.06396\,\Omega$$

The resistance of a cable is given by

$$R = \frac{\rho l}{A}$$

where ρ is the resistivity (Ωm)
   l is the length (m)
   A is the cross-sectional area (m^2)

so that

$$A = \frac{\rho l}{R}$$

$$A = \frac{1.78\,\Omega\mathrm{m} \times 120\,\mathrm{m}}{10^8 \times 0.06396\,\Omega}$$

$$= \frac{0.33}{10^4}\,\mathrm{m}^2$$

$$= \frac{0.33}{10^4}\,\cancel{\mathrm{m}^2}\left[\frac{10^6\,\mathrm{mm}^2}{1\,\cancel{\mathrm{m}^2}}\right]$$

$$= \underline{33\,\mathrm{mm}^2}$$

EXAMPLE 4 The nearest standard size cable to that required for example C is $35\,\mathrm{mm}^2$. Calculate the actual voltage drop using this cable and the line voltage at the load end.

We notice that the resistance of a cable is inversely proportional to its cross-sectional area, i.e.

$$R \propto \frac{1}{A}$$

and also that the voltage drop for a given current is directly proportional to the resistance,

i.e. $U \propto R$

so that $U \propto R \propto \dfrac{1}{A}$

or $U \propto \dfrac{1}{A}$

Thus if U_2 is the voltage drop when the area is A_2 and U_1 is the voltage drop when the area is A_1

then $U_2 \propto \dfrac{1}{A_2}$ or $U_2 = \dfrac{k}{A_2}$

and $U_1 \propto \dfrac{1}{A_1}$ or $U_1 = \dfrac{k}{A_1}$ where k is a constant.

Dividing we have

$$\frac{U_2}{U_1} = \frac{A_1}{A_2} \quad k \text{ cancelling.}$$

Let $U_1 = 5.77$ when $A_1 = 33\,\mathrm{mm}^2$ as calculated and U_2 be the new voltage when $A_2 = 35\,\mathrm{mm}^2$

$$\frac{U_2}{5.77} = \frac{33}{35}$$

$$U_2 = 5.77 \times \frac{33}{35}$$

$$= 5.44\,\text{V}$$

The line voltage at load end

$$= 400 - \sqrt{3} \times 5.44$$

$$= 400 - 9.42$$

$$= 390.58$$

$$\text{or } \underline{391\,\text{V}}$$

Again, this is not a rigid treatment but the result is sufficiently accurate for many practical purposes.

EXAMPLE 5 A balanced load of 25 A is to be supplied from the 400 V mains to a point 150 m away. The total voltage drop must not exceed 4% of the supply voltage. Choose the most suitable cable from those given below.

cable size (mm^2)	6	10	16	25
current rating (A)	34	46	62	80
voltage drop per ampère per metre (mV)	6.4	3.8	2.4	1.5

Permissible total voltage drop

$$= 4.0\% \times 400$$

$$= 16\,\text{V}$$

Denoting the figures in the bottom line of the table by x we have

$$x\frac{\text{mV}}{\text{A.m}}\left[\frac{1\,\text{V}}{1000\,\text{mV}}\right] \times 25\,\text{A} \times 150\,\text{m} \leqslant 16$$

$$\frac{x \times 25 \times 150}{1000} \leqslant 16 \quad (\leqslant \text{means equal to or less than})$$

$$x \leqslant \frac{16 \times 1000}{25 \times 150}$$

$$\leqslant 4.27$$

so we choose a figure in the bottom line of the table which is less than 4.27.

The appropriate size of cable is thus $10 \, \text{mm}^2$.

The procedure then is to choose a cable having a voltage drop figure which is less than

$$\frac{\text{permissible voltage drop} \times 1000}{\text{current required} \times \text{length of run (metres)}}$$

For the following examples reference to BS 7671 and the *IEE On-Site Guide* will be necessary.

EXAMPLE 6 A p.v.c. trunking is to be used to enclose single-core p.v.c-insulated distribution cables (copper conductors) for a distance of 30 m from the main switchgear of an office building to supply a new 400 V T.P. and N distribution fuseboard. The balanced load consists of 24 kW of discharge lighting. The fuses at the main switch-fuse and at the distribution board are to BS 88 part 2. The voltage drop in the cables must not exceed 6 V. The ambient temperature is anticipated to be 35 °C. The declared value of I_p is 20 kA and that of Z_e is 0.30 Ω. Assume that the requirements of BS 7671 434-03 are satisfied by the use of BS 88 fuses.

(a) For the distribution cables, establish the:
- (i)   design current (I_b)
- (ii) minimum rating of fuse in the main switch-fuse (I_n)
- (iii) maximum mV/A/m value
- (iv) minimum current rating (I_t)
- (v)   minimum cross-sectional area of the live conductors
- (vi) actual voltage drop in the cables.

(b) It is proposed to install a 2.5 mm² protective conductor within the p.v.c. trunking. Verify that this meets shock protection requirements.   (C & G)

(a)(i) Design current $I_b = \dfrac{24 \times 10^3 \times 1.8}{\sqrt{3} \times 400}$ (1.8 factor for discharge lighting)

$$= 62.36 \, \text{A}$$

(ii) Minimum BS 88 fuse rating (I_n) is 63 A.

(iii) Maximum mV/A/m value $= \dfrac{6 \times 1000}{62.36 \times 30}$

$$= 3.2 \, \text{mV/A/m}$$

(iv) Minimum current rating $(I_t) = \dfrac{63}{0.94}$ (temperature correction factor C_a for $35\,°\text{C}$)

$$= 67.02 \, \text{A}$$

(v) Minimum c.s.a. of cable is $16 \, \text{mm}^2$ ($68\,\text{A}$ $2.4\,\text{mV/A/m}$).

(vi) Actual voltage drop in $30\,\text{m} = \dfrac{2.4 \times 62.36 \times 30}{1000}$

$$= 4.49 \, \text{V}$$

(b) Check compliance with Table 41D (BS 7671) using *IEE On-site Guide*.

From Table 9A, $R_1 + R_2$ for $16\,\text{mm}^2/2.5\,\text{mm}^2 = 1.15 + 7.41 \, \text{m}\Omega/\text{m}$.

From Table 9C, factor of 1.20 must be applied.

Now $Z_s = Z_e + R_1 + R_2$

$$R_1 + R_2 = \frac{30 \times (1.15 + 7.41) \times 1.20}{1000}$$

$$= 0.308 \, \Omega$$

$$\therefore \quad Z_s = 0.3 + 0.308$$

$$Z_s = 0.608 \, \Omega$$

This satisfies Table 41D as the maximum Z_s for a $63\,\text{A}$ fuse is $0.86\,\Omega$.

EXAMPLE 7 Figure 46 shows the layout of a factory conveyor system employing twelve 230 V, 460 W heaters placed 5 m apart used for drying products. The heaters are arranged in three

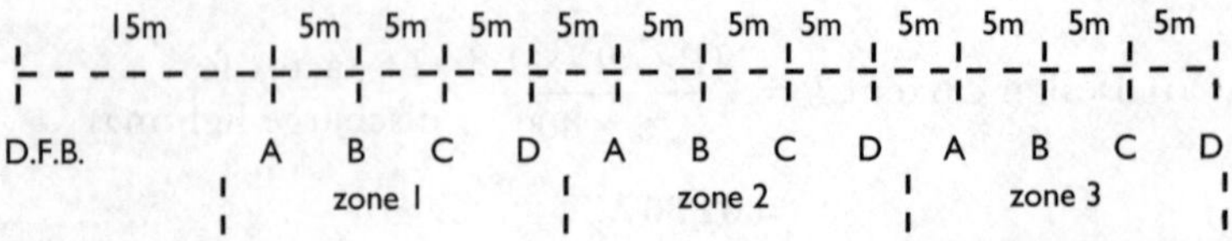

Fig. 46

distinct zones 1, 2, and 3. Each zone is fed by a separate phase from a T.P. and N distribution board containing BS 88 fuses.

The distance from the distribution board to the first heater is 15 m. A maximum permitted voltage drop of 6 V between the distribution board and each heater is specified.

The supply to the heaters is to be by single-core p.v.c. insulated cables together in a single conduit along with a single communal $4\,mm^2$ circuit protective conductor. An ambient temperature of 45 °C may be assumed.

(a) Redraw the layout and indicate the current in each section.
(b) Determine the:
    (i)   minimum fuse ratings in the distribution board
    (ii) minimum cable current rating
    (iii) minimum cable cross-sectional area for each zone
    (iv) actual voltage at the final heater (D) of *each* zone.

(C & G) (modified)

(a) See Fig. 47.

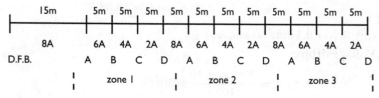

Fig. 47

(b) Load current for each heater is $\dfrac{460}{230} = 2\,A$.
Thus each zone requires 8 A (I_b).

(i) Fuse rating (I_n) for each zone is 10 A.

(ii) Zone 1.

Three circuits so C_g is 0.7 and 45 °C so C_a is 0.79,

$$\text{so minimum cable current rating} = \frac{10}{0.7 \times 0.79}$$

$$I_t = 18.08\,A$$

(iii) Select from Table 4D1A (BS 7671) or Table 6D1 (*IEE On-site Guide*) $2.5\,mm^2$ cable (24 A). Identify from Table 4D1B (BS 7671)

75

or Table 6D2 (*IEE On-site Guide*) mV/A/m value for $2.5\,\text{mm}^2$ as $18\,\text{mV/A/m}$.

Voltage drops:

$$\text{D.F.B. to A} = \frac{8 \times 15 \times 18}{1000} = 2.16\,\text{V}$$

$$\text{A to B} \quad = \frac{6 \times 5 \times 18}{1000} = 0.54\,\text{V}$$

$$\text{B to C} \quad = \frac{4 \times 5 \times 18}{1000} = 0.36\,\text{V}$$

$$\text{C to D} \quad = \frac{2 \times 5 \times 18}{1000} = 0.18\,\text{V}$$

(iv) Voltage at heater D $= 230 - 2.16 - 0.54 - 0.36 - 0.18$
$$= 225.3\,\text{V}$$

(ii) Zone 2.

Two circuits so C_g is 0.8 and $45\,^\circ\text{C}$ so C_a is 0.79 but the first conduit section determines cable rating,

$$\text{so minimum cable current rating} = \frac{10}{0.7 \times 0.79}$$
$$I_t = 18.08\,\text{A}$$

(iii) Select $2.5\,\text{mm}^2$ as above.

Voltage drops:

$$\text{D.F.B. to A} = \frac{8 \times 35 \times 18}{1000} = 5.04\,\text{V}$$

$$\text{A to B} \quad = \frac{6 \times 5 \times 18}{1000} = 0.54\,\text{V}$$

$$\text{D.F.B. to C} = \frac{4 \times 5 \times 18}{1000} = 0.36\,\text{V}$$

$$\text{D.F.B. to D} = \frac{2 \times 5 \times 18}{1000} = 0.18\,\text{V}$$

(iv) Voltage at heater D $= 230 - 5.04 - 0.54 - 0.36 - 0.18$
$$= 223.88\,\text{V}$$

(ii) Zone 3.

(iii) Minimum cable c.s.a. is $2.5\,\text{mm}^2$ as above.

Voltage drops:

$$\text{D.F.B. to A} = \frac{8 \times 55 \times 18}{1000} = 7.92\,\text{V}$$

(iii) This is too high so go to $4\,mm^2$ (32 A, 11 mV/A/m)

$$\text{D.F.B. to A} = \frac{8 \times 55 \times 11}{1000} = 4.84\,V$$

$$\text{A to B} \quad = \frac{6 \times 5 \times 11}{1000} = 0.33\,V$$

$$\text{B to C} \quad = \frac{4 \times 5 \times 11}{1000} = 0.22\,V$$

$$\text{C to D} \quad = \frac{2 \times 5 \times 11}{1000} = 0.11\,V$$

(iv) Voltage at heater D $= 230 - 4.84 - 0.33 - 0.22 - 0.11$

$$= 224.5\,V$$

EXAMPLE 8 It is proposed to install a new 230 V 50 Hz distribution board in a factory kitchen some 40 m distant from the supplier's intake position.

It is to be supplied by two $25\,mm^2$ p.v.c. insulated (copper conductors) single-core cables in steel conduit. Protection at origin of the cables is to be by BS 88 fuses rated at 80 A.

It is necessary for contractual purposes to establish:
(a) the prospective short circuit current (p.s.c.c.) at the distribution board, and
(b) that the proposed distribution cables will comply with BS 7671 requirements 434-03-03.
A test conducted at the intake position between phase and neutral to determine the external impedance of the supplier's system indicates a value of $0.12\,\Omega$.

(a) The resistance of distribution cables from intake to distribution board
From Table 9A (*IEE On-site Guide*), R_1/R_2 for $25\,mm^2/25\,mm^2$ cables $= 1.454\,mV/m$.
From Table 9C a multiplier of 1.20 is necessary using the Table 6A figures as

$$R_1/R_2 = \frac{40 \times 1.454 \times 1.20}{1000}$$

$$= 0.0698\,\Omega \text{ (regard this as impedance)}$$

So total short circuit fault impedance $= 0.12 + 0.0698$

$$= 0.19\,\Omega$$

Thus $\qquad I_f = \dfrac{230}{0.19}$

$\therefore \qquad$ p.s.c.c. $= 1210\,\mathrm{A}$

From Appendix 3, Fig. 3.3A, the BS 88 fuse clearance time is approximately 0.1 s.

(b) From Requirement 434-03-03, $t = \dfrac{k^2 S^2}{I^2}$

$$= \dfrac{115^2 \times 25^2}{1200^2}$$

$\therefore$ limiting time for conductors $(t) = 5.74\,\mathrm{s}$

The cables are disconnected well before the $25\,\mathrm{mm}^2$ cable conductors reach their limiting temperature, thus they are protected thermally.

EXAMPLE 9 Two $25\,\mathrm{mm}^2$ single-core p.v.c. insulated cables (copper conductors) are drawn into a p.v.c. conduit along with a $10\,\mathrm{mm}^2$ protective conductor to feed a 230 V industrial heater.

The following details are relevant:

Protection at the origin is by 80 A BS 88 fuses.

The tested value of Z_e at the cables origin is $0.35\,\Omega$.

The length of cables run is 55 m.

(a) Establish the:
   (i)   value of $R_1 + R_2$ of the cables
   (ii) prospective earth fault loop current (I_{ef})
   (iii) the disconnection time of the fuse.

(b) Does the clearance time comply with BS 7671?

(a)(i) Using the *IEE On-site Guide*
From Table 9A $R_1 + R_2$ for $25\,\mathrm{mm}^2/10\,\mathrm{mm}^2$ cables $= 2.557\,\mathrm{m}\Omega/\mathrm{m}$.
From Table 9C apply the factor 1.20

$$\text{Thus } R_1 + R_2 = \dfrac{55 \times 2.557 \times 1.20}{1000}$$

$$= 0.169\,\Omega$$

So Z_s at distribution board $= 0.35 + 0.169$

$$= 0.519\,\Omega$$

(ii) Prospective earth fault current $(I_{ef}) = \dfrac{230}{0.519}$

$$= 443\,\text{A}$$

(iii) Using BS 7671, Appendix 3, Table 3.3A, disconnection time is 3.8 s.

(b) The clearance time complies with BS 7671 Requirement 413-02-13 which specifies a maximum disconnection time of 5 s.

EXAMPLE 10 It is necessary to confirm that the cross-sectional area of the protective conductor in a previously installed 400/230 V distribution circuit complies with BS 7671 requirement 543-01-03. The phase conductors are 10 mm^2 and the circuit-protective conductor is 2.5 mm^2. The length of the cables run in plastic conduit is 85 m. Protection is by 32 A, BS 88 fuses and the value of Z_e is 0.4 Ω.

Using the *IEE On-site Guide*

From Table 9A $R_1 + R_2$ for 10 mm^2/2.5 mm^2 cables
$= 1.83 + 7.41$ mΩ/m

From Table 9C apply the factor 1.20

Thus $R_1 + R_2 = \dfrac{85 \times (1.83 + 7.41) \times 1.20}{1000}$

$$= 0.942\,\Omega$$

So Z_s at distribution board $= 0.4 + 0.942$

$$= 1.342\,\Omega$$

Prospective earth fault loop current $(I_{ef}) = \dfrac{230}{1.342}$

$$= 171\,\text{A}$$

Using BS 7671

From Appendix 3 Table 3.3A the fuse clearance time is 0.9 s.

From BS 7671 requirement 543-01-03

$$s = \frac{\sqrt{I^2 t}}{k} \quad (k \text{ is } 115 \text{ Table } 54C)$$

$$= \frac{\sqrt{171^2 \times 0.9}}{115}$$

$$= \underline{1.41 \text{ mm}^2}$$

This confirms that a 2.5 mm² protective conductor is acceptable.

EXAMPLE 11 The declared value of I_p at the origin of a 230 V 50 Hz installation is 1.5 kA. The length of 25 mm² p.v.c./p.v.c. meter tails is 2 m; at this point a switch-fuse containing 100 A BS 88 Part 2 fuses is to be installed to provide control and protection for a new installation. A 20 m length of 16 mm² heavy duty mineral insulated cable (exposed to touch), (copper conductors and sheath) is to be run from the switch-fuse to a new distribution board.

(a) Establish that the mineral cable complies with BS 7671 requirement 434-03.

(b) How could you ensure that the requirements of BS 7671 413-02-01, etc. and Chapter 7 are satisfied?

(a) Resistance of 2 m of 25 mm² meter tails using mV/A/m value from Table 9D1A as ohms per metre at 70 °C

$$R_{mt} = 2 \times 0.00175$$

$$= 0.0035 \, \Omega$$

Resistance of 20 m of 16 mm² twin m.i.c.c. cable using mV/A/m values from Table 4JB as ohms per metre at 70 °C

$$R_{mi} = 20 \times 0.0026$$

$$= 0.052 \, \Omega$$

Impedance of supply $= \dfrac{230}{1500}$

$$= 0.153 \, \Omega$$

Thus total impedance from source to distribution board

$$= 0.153 + 0.0035 + 0.052 = 0.2085 \, \Omega$$

Prospective short circuit fault current

$$I_p = \frac{230}{0.2085} = 1103 \, \text{A}$$

Disconnection time from BS 7671 Table 3.3B is 0.3 s

Now using the 434-03-03 adiabatic equation

$$t = \frac{k^2 S^2}{I^2}$$

$$= \frac{135^2 \times 16^2}{1103^2}$$

$$= 3.8 \, \text{s}$$

Thus 16 mm^2 cable is protected against the thermal deterioration.

(b) As no details are available in BS 7671 in relation to the resistance/impedance of the m.i.c.c. sheath, the prospective value of Z_s could not be established, but the actual value must be tested when the installation is commissioned and the value recorded in the Electrical Installation Certificate referred to in requirement 742.

EXAMPLE 12 A 230 V, 50 Hz, 5 kW electric motor is fed from a distribution board containing BS 88 Part 2 fuses. The wiring between the d.f.b. and the motor starter which is 20 m distant is p.v.c. insulated single-core cables drawn into steel conduit. Assume that the:

(i)   starter affords overload protection;

(ii)   motor has a power factor of 0.75 and an efficiency of 80%;

(iii) ambient temperature is 40 °C;

(iv) fuse in the d.f.b. may have a rating up to twice the rating of the circuit cables;

(v)   volts drop in the motor circuit cables must not exceed 6 V;

(vi) resistance of metal conduit is 0.1 Ω per metre;

(vii) 'worst' conduit run is 8 m with two 90° bends;

(viii) I_p at d.f.b. is 2 kA;

(ix)   value of Z_e is $0.19\,\Omega$.

Establish the:

(a) design current (I_b);

(b) rating of circuit fuse;

(c) minimum cable rating (I_n) between d.f.b. and starter;

(d) minimum cable cross-sectional area;

(e) actual voltage drop in cables;

(f) prospective short circuit current;

(g) short circuit disconnection time;

(h) whether BS 7671 requirement 434-03-03, etc. is satisfied;

(i) whether BS 7671 requirement 413-02-04, etc. is satisfied;

(j) minimum conduit size.

(a)   Design current $(I_b) = \dfrac{5000}{230 \times 0.75 \times 0.8}$

$$= 36.2\,\text{A}$$

(b) Rating of circuit fuse may be 80 A.

(c) Minimum cable rating may be 40 A.

(d) Minimum cable c.s.a. $= \dfrac{40}{0.87} = 46\,\text{A}$

from Table 4D1A select $10\,\text{mm}^2$ cable (57 A).

(e) Actual voltage drop:

from Table 4D1B mV/A/m value for $10\,\text{mm}^2$ is 4.4 thus

$$\text{volts drop} = \frac{36.2 \times 20 \times 4.4}{1000}$$

$$= 3.19\,\text{V}$$

(f) Impedance of supply cables to d.f.b.

$$= \frac{230}{2000} = 0.115\,\Omega$$

Using BS 7671 Tables 9A and 9C, resistance of circuit cables

$$= \frac{20 \times 3.66 \times 1.2}{1000}$$

$$= 0.09\,\Omega$$

thus total circuit impedance $= 0.115 + 0.09 = 0.205\,\Omega$

Prospective short circuit current $= \dfrac{230}{0.205}$

$$= 1122\,\text{A}$$

(g) Disconnection time from Fig. 3.3A is 0.1 s

(h) Cable thermal capacity $t = \dfrac{k^2 \times S^2}{I^2}$

$$= \dfrac{115^2 \times 10^2}{1122^2}$$

$$= 1.05\,\text{s}$$

Thus 10 mm^2 cables are thermally safe.

(i) Now $Z_s = Z_e + R_1 + R_2$

Using BS 7671 Tables 9A and 9C,

Resistance of $R_1 \qquad = \dfrac{20 \times 1.83}{1000} = 0.0366\,\Omega$

Resistance of conduit $R_2 = 20 \times 0.01 = 0.2\,\Omega$

Thus $Z_s = 0.19 + 0.0366 + 0.2 = 0.4266\,\Omega$

$$I_{ef} = \dfrac{230}{0.4266} = 539\,\text{A}$$

From BS 7671, Fig. 3.3A, disconnection time is 1.4 s; this being less than 5 s, protection is satisfactory.

(j) From Table 5C,

cable factor for $2 \times 10\,\text{mm}^2$ cables $= 2 \times 105 = 210$

From Table 5D select 25 mm conduit (factor 292).

EXERCISE 7

1. A balanced load of 30 A is supplied through a cable each core of which has resistance 0.28 Ω. The line voltage at the supply end is 400 V. Calculate the voltage at the load end, the percentage total voltage drop and the power wasted in the cable.

2. Each core of a three-core cable, 164 m long, has a cross-sectional area of 35 mm^2. The cable supplies power to a 30 kW, 400 V, three-phase motor working at full load with 87% efficiency and power factor 0.72 lagging.

Calculate:

(a) the voltage required at the supply end of the cable;

(b) the power loss in the cable.

The resistivity of copper may be taken as $1.78 \times 10^{-8}\ \Omega\text{m}$ and the reactance of the cable may be neglected.

3. A 40 kW, 400 V, three-phase motor, running at full load, has efficiency 86% and power factor 0.75 lagging. The three-core cable connecting the motor to the switchboard is 110 m long and its conductors are of copper 25 mm^2 in cross-section.

   Calculate the total voltage drop in the cable, neglecting reactance.

   If the cable runs underground for most of its length, choose a suitable type of cable for the purpose and give a descriptive sketch of the system of laying it.

   The resistivity of copper may be taken as $1.78 \times 10^{-8}\ \Omega\text{m}$.

4. The estimated load in a factory extension is 200 kW at 0.85 p.f. (balanced). The supply point is 75 m away where the line voltage is 400 V. Choose the most suitable size of cable from those given below in order that the total voltage drop shall not exceed 2.5% of supply voltage.

   Cross-sectional areas
   of available conductors (mm^2)   35   50   70   95
   (Resistivity of conductor is $1.78 \times 10^{-8}\ \Omega\text{m}$.)

5. A motor taking 200 kW at 0.76 p.f. is supplied at 400 V three-phase by means of a three-core copper cable 200 m long.

   (a) Calculate the minimum cable cross-sectional area if the voltage drop is not to exceed 5 V.

   (b) If the cable size calculated is non-standard, select from the table a suitable standard cable and calculate the actual voltage drop using that cable.

   Standard cross-sectional areas of cable conductors (mm^2)
       300   400   500   630
   (Resistivity of copper $1.78 \times 10^{-8}\ \Omega\text{m}$.)

6. A three-phase current of 35 A is supplied to a point 75 m away by a cable which produces a voltage drop of 2.2 mV per ampere per metre. Calculate the total voltage drop.

   The following questions should be answered by reference to the appropriate tables in BS 7671 and/or in the *IEE On-site Guide* to BS 7671.

7. A balanced load of 85 A is required at a point 250 m distant from a 400 V supply position. Choose a suitable cable (clipped direct) from Tables 4E4A and 4E4B in order that the total voltage drop shall be within the BS 7671 specified limit (ambient temperature 30 °C).

8. A 25 kW, 400 V three-phase motor having full load efficiency and power factor 80% and 0.85 respectively is supplied from a point 160 m away from the main switchboard. It is intended to employ a surface-run, multicore p.v.c.-insulated cable, non-armoured (copper conductors). The ambient temperature is 30 °C and BS 88 fuses are to be employed at the main switchboard. Select a cable to satisfy the BS 7671 requirements.

9. The total load on a factory sub-distribution board consists of:
10 kW lighting balanced over three phases, unity power factor;
50 kW heating balanced over three phases, unity power factor and
30 kW motor load having an efficiency 80%, power factor 0.8.
The line voltage is 400 V and the supply point is 130 m distant.
Protection at the origin of the cable (clipped direct) is by BS 88 fuses.
The ambient temperature is 30 °C.
Select a suitable cable from Tables 4D2A and 4D2B, in order that the voltage drop shall not exceed 3% of the supply voltage.

10. Calculate the additional load in amperes which could be supplied by the cable chosen for question 9 with the voltage drop remaining within the specified limits.

11. A 12 kW, 400 V three-phase industrial heater is to be wired using single-core p.v.c. insulated cables (copper conductors) 30 m in length drawn into a steel conduit. The following details may be relevant to your calculation.
Ambient temperature 40 °C.
Protection by BS 3036 (semi-enclosed) fuses.
Voltage drop in the cables must not exceed 10 V.
The contract document calls for a 2.5 mm² conductor to be drawn into the conduit as a supplementary protective conductor.
The worst section of the conduit run involves two right-angle bends in 7 m. Establish the:
(a) design current (I_b);
(b) minimum fuse rating (I_n);
(c) maximum mV/A/m value;

(d) minimum live cable rating (I_t);

   (e) minimum live cable c.s.a.;

   (f) actual voltage drop;

   (g) minimum conduit size.

12. The external live conductor impedance and external earth fault loop impedance are tested at the intake of a 230 V single-phase installation and show values of $0.14\,\Omega$ and $0.28\,\Omega$ respectively. A p.v.c. trunking runs from the intake position to a distribution board 40 m distant and contains 35 mm^2 live conductors and a 10 mm^2 protective conductor.

   (a) Estimate the:

   (i) prospective short circuit current (p.s.c.c.) at the distribution board;

   (ii) p.s.c.c. clearance time of the 100 A BS 88 fuse at the origin of the cable;

   (iii) value of the earth fault loop impedance (Z_s) at the distribution board;

   (iv) prospective earth fault loop current;

   (v) earth fault clearance time of the BS 88 fuse at the origin of the cable.

   (b) State the maximum permitted value of Z_s under these conditions.

13. A p.v.c. trunking containing single-core p.v.c.-insulated distribution cables (copper conductors) is to be run 30 m from the 400/230 V main switchgear of an office building to supply a new T.P. and N distribution fuseboard. The balanced load consists of 24 kW of discharge lighting. The fuses at the main switch-fuse and at the distribution board are to BS 88 part 2. The voltage drop in the distribution cables must not exceed 6 V and the ambient temperature is anticipated to be 35 °C. The declared value of I_p is 20 kA and that of Z_e is 0.3 Ω. Assume that the requirements of BS 7671 434-5 are satisfied.

   (a) For the distribution cables, establish and state the

   (i) design current;

   (ii) minimum rating of fuse in the main switch fuse;

   (iii) maximum mV/A/m value;

   (iv) minimum current rating;

(v) minimum cross-sectional area of the live conductors;

(vi) actual voltage drop in the cable.

(b) It is proposed to install a $4\,mm^2$ protective conductor within the p.v.c. trunking.

(i) State the value of Z_s.

(ii) Verify that this meets BS 7671 shock protection requirements.

14. A security building is to be built at the entrance to a factory. This new building is to be provided with a 230 V single-phase supply and is to be situated 20 m from the main switchroom. A 30 m twin p.v.c.-insulated armoured underground cable (copper conductors) supplies the new building, which allows 5 m at each end for runs within the main switchroom and security buildings. The connected load in the security building comprises

   one 3 kW convector heater

   two 1 kW radiators

   two 1.5 kW water heaters (instantaneous type)

   one 6 kW cooker

   six 13 A socket outlets (ring circuit)

   a 2 kW lighting load.

   Diversity may be applied (business premises).

   Establish:

   (a) the prospective maximum demand;

   (b) minimum current rating of the switch-fuse in the switchroom at the origin of the underground cable.

   (c) Determine

   (i) the minimum current rating of p.v.c.-insulated twin armoured (copper conductor) underground cable, assuming an ambient temperature of 25 °C and protection to be by BS 88 part 2 devices;

   (ii) the minimum size of cable, assuming the voltage drop is limited to 2 V;

   (iii) the actual voltage drop in the cable.   (C & G)

15. It is proposed to install a p.v.c.-insulated armoured cable to feed a 25 kW, 400 V, three-phase 50 Hz resistive element type of furnace. The cable is to be surface run along a brick wall of a factory and has a total length of 95 m. The protection at the origin of the circuit is to be

by BS 88 fuses. The cable armour may be relied upon as the circuit protective conductor. The ambient temperature in the factory will not exceed 35 °C and the voltage drop must not exceed 10 V.

Determine and state the:

(a) design current;

(b) fuse rating;

(c) minimum cable current rating;

(d) maximum mV/A/m value;

(e) minimum cross-sectional area of the live conductors;

(f) actual voltage drop in the cable. (C & G)

16. A pipeline pump is connected to a 400/230 V three-phase supply. It is wired in 1.5 mm^2 p.v.c.-insulated cables drawn into 25 mm steel conduit running 30 m from a distribution board containing BS 88 part 2 fuses (10 A fuses protect the pump).

It is now necessary to add alongside the pump a 12 kW 400/230 V heater and it is proposed to draw the p.v.c.-insulated cables into the existing 25 mm pump circuit conduit and insert suitable fuses into the distribution board which has vacant ways.

The following assumptions may be made:

(i) an ambient temperature of 35 °C;

(ii) the maximum distance between draw in boxes is 9.5 m with two right-angle bends;

(iii) the maximum voltage drop in the heater circuit is 5 V;

(iv) a 2.5 mm^2 protective conductor is installed in the steel conduit to satisfy a clause in the electrical specifications.

Determine for the heater circuit the:

(a) design current I_b;

(b) suitable fuse rating I_n;

(c) maximum mV/A/m value;

(d) minimum cable current rating I_t;

(e) minimum cable c.s.a.;

(f) actual voltage drop.

For the existing pump circuit establish whether the:

(g) pump circuit cable current rating is still adequate;

(h) 25 mm conduit is suitable for the additional cables.

Electromagnetism I

References: **Magnetic flux and flux density.**
Magnetizing force.
Permeability and relative permeability.
Series and parallel magnetic circuits.

MAGNETIZING FORCE (*H*)

To find the magnetizing force of a coil having N turns and carrying a current I ampere when it is wound on a magnetic circuit of mean length l m.

$$H = \frac{I \times N}{l} \text{ ampère turns per metre (At/m)}$$

The product $I \times N$ is called the magnetomotive force (m.m.f.) of the coil.

EXAMPLE I   Determine the magnetizing force of a coil of 100 turns, carrying a current of three amperes, on a magnetic circuit 150 mm long.

$$H = \frac{I \times N}{l}$$

$$= \frac{3 \times 100}{150/1000} \quad \text{(Note conversion of 150 mm to metres)}$$

$$= \frac{3 \times 100 \times 1000}{150}$$

$$= \underline{2000 \text{ ampere turns/metre}}$$

EXAMPLE 2   The air gap in a magnetic circuit is 1.0 mm long. The magnetizing force required to set up a certain value of flux in this gap is found to be 200 000 ampere turns per metre.

Calculate:

(a) the number of ampere turns required for the gap;

(b) the current required if the circuit is energized by a coil of 1000 turns.

(a)
$$H = \frac{I \times N}{l}$$

$$200\,000 = \frac{I \times N}{1.0/1000}$$

In this case l is the length of the air gap

$$\therefore \quad I \times N = 200\,000 \times \frac{1.0}{1000}$$

$$= \underline{200 \text{ ampere turns}}$$

(b)
$$I \times N = 200$$

$$I \times 1000 = 200$$

$$\therefore \quad I = \frac{200}{1000}$$

$$= \underline{0.2\,\text{A}}$$

PERMEABILITY

In air, $\quad B = H \times \mu_0$

and $\quad \mu_0 = 4\pi \times 10^{-7}$,

where μ_0 is the permeability of free space.

EXAMPLE 3 Calculate the flux density produced in air by a magnetizing force of 20 000 At/m (ampere turns per metre).

$$B = H \times \mu_0$$

$$= 20\,000 \times 4\pi \times 10^{-7}$$

$$= \frac{20\,000 \times 4\pi}{10\,000\,000}$$

$$= \frac{25.1 \times 10\,000}{10\,000\,000}$$

$$= \underline{0.025 \text{ Tesla (T) or Wb/m}^2}$$

EXAMPLE 4 Determine the magnetizing force required to produce a flux density of 1.2 mT in air.

$$B = H \times \mu_0$$

$$\frac{1.2}{1000} = \frac{4\pi}{10^7} \times H$$

(note the conversion of mT to T)

$$\therefore \quad H = \frac{1.2 \times 10\,000\,000}{4 \times \pi \times 1000}$$

$$= \underline{955\,\text{At/m}}$$

EXAMPLE 5 The air gap in a certain machine is $0.5\,\text{m}^2$ in cross-section and is 5 mm long. The magnetizing force is provided by a coil of 1000 turns. Calculate the current which must flow in the coil to produce a total flux of 0.25 Wb in the gap.

Find the flux density:

$$\Phi = B \times A$$

$$0.25 = B \times 0.5$$

$$\therefore \quad B = \frac{0.25}{0.5}$$

$$= 0.5\,\text{T}$$

Find the magnetizing force:

$$B = H \times \mu_0$$

$$0.5 = H \times 4\pi \times 10^{-7}$$

$$\therefore \quad H = \frac{0.5}{4\pi \times 10^{-7}}$$

$$= \frac{0.5 \times 10\,000\,000}{4\pi}$$

$$= 398\,000\,\text{At/m}$$

Ampere turns required for gap:

$$H = \frac{I \times N}{l}$$

$$398\,000 = \frac{I \times N}{5/1000}$$

$$\therefore \quad I \times N = 398\,000 \times \frac{5}{1000}$$

$$= 1990$$

Find the current:

$$I \times N = 1990$$

$$I \times 1000 = 1990$$

$$\therefore \quad I = \frac{1990}{1000}$$

$$= \underline{1.99\,\text{A}}$$

RELATIVE PERMEABILITY

In magnetic materials

$$B = H \times \mu_0 \times \mu_r$$

where μ_r is the relative permeability of the steel.

EXAMPLE 6 Calculate the flux density produced by a magnetizing force of 2000 At/m in a steel of relative permeability 400.

$$B = H \times \mu_0 \times \mu_r$$

$$= \frac{2000 \times 4\pi \times 400}{10\,000\,000}$$

$$= \underline{1.01\,\text{T}}$$

EXAMPLE 7 A mild steel ring 200 mm² in cross-sectional area and 0.2 m in mean diameter is wound with 450 turns of wire. Assuming that the steel has a relative permeability of 600, calculate the total flux produced in the ring when a current of 2 A flows in the coil.

Find the magnetizing force:

$$H = \frac{I \times N}{l}$$

(in this case l is the mean circumference of the ring)

$$= \frac{2 \times 450}{0.2 \times \pi}$$

$$= 1430 \,\text{At/m}$$

Find the flux density produced:

$$B = H \times \mu_0 \times \mu_r$$

$$= \frac{1430 \times 4 \times \pi \times 600}{10\,000\,000}$$

$$= \underline{1.08\,\text{T}}$$

Find the total flux:

$$\Phi = B \times A$$

$$= 1.08 \times \frac{200}{1\,000\,000}$$

$$= \underline{0.000216\,\text{Wb}}$$

$$= \underline{0.216\,\text{mWb}}$$

EXAMPLE 8 The relationship between the flux density and the magnetizing force required in a certain brand of steel is

B (T)	0.8	1.0	1.2
H At/m	240	400	650

A ring is formed from this steel 50 mm^2 in cross-section and 0.1 m in mean diameter. A coil of 2500 turns is wound evenly upon it. Calculate the current which must flow in the coil to magnetize the steel to a flux density of 1.1 T.

A graph to show the relationship between flux density and magnetizing force can be plotted (see Fig. 48) from the information given.

From the graph the magnetizing force required to set up a flux density of 1.1 T in this steel is 500 At/m.

$$H = \frac{I \times N}{l}$$

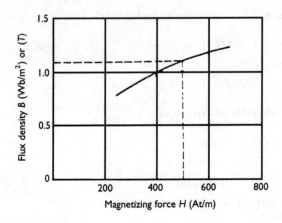

Fig. 48

m.m.f. required:

$$I \times N = H \times l$$
$$= 500 \times 0.1\pi$$
$$= 50\pi \text{ At}$$
$$I = \frac{50\pi}{2500} \frac{A\cancel{t}}{\cancel{t}}$$
$$= \underline{0.0628 \text{ A}}$$

EXAMPLE 9 If a sawcut 1 mm wide is now made through the ring of Example 8 as shown in Fig. 49, determine the current which now must flow in order to produce the same flux density as before.

This is now an example of a series type magnetic circuit. The same general principles apply to it as apply in series electric circuits. That is:

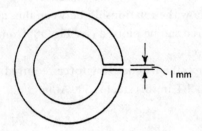

Fig. 49

(1) the same *total* flux is assumed to exist at each point;

(2) the total m.m.f. required is found by adding the m.m.fs. required to magnetize the separate sections.

It is assumed that the 1 mm gap does not effectively change the length of the iron part of the circuit; this remains as 0.1π m. The m.m.f. for the steel portion is thus

$$50\pi = 157.1\,\text{At as before}$$

For the gap

$$\frac{B}{H} = \mu_0$$

$$\frac{1.1}{H} = 4\pi \times 10^{-7}$$

$$H = \frac{1.1}{4\pi \times 10^{-7}}$$

$$= \frac{1.1}{4\pi} \times 10^7$$

$$= 875\,200\,\text{At/m}$$

$$\text{m.m.f.} = H \times l$$

$$\text{m.m.f.} = 875\,200 \times \frac{1}{1000}$$

(here l is the length of the gap, and note the conversion to metres)

$$= 875.2\,\text{At}$$

The total m.m.f. required

$$= 157.1 + 875.2$$

$$= 1032.3\,\text{At}$$

and the magnetizing current

$$= \frac{1032.3}{2500}\,\frac{\text{A}l}{l}$$

$$= \underline{0.413\,\text{A}}$$

No allowance has been made for the effects of leakage and fringing but this magnetic circuit is typical of that found in many types of relay and transformer core.

EXAMPLE 10 A magnetic circuit has the form shown in Fig. 50. The material is the same as that used in Examples 8 and 9. The centre limb is produced with a winding of 1000 turns. Determine the current required in this winding to produce a total flux in the air gap of 0.12 mWb.

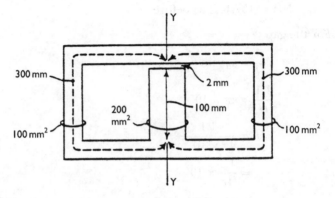

Fig. 50

It will be noticed that the magnetic circuit is symmetrical about the line YY; that is, the outside limbs are identical in every respect. This is typical of many magnetic circuits such as are found in relay and transformer cores, d.c. machines, etc. If the effects of leakage and fringing are ignored, the circuit may be folded about the line YY so forming the equivalent series circuit shown in Fig. 51.

We now proceed as in Examples 8 and 9.

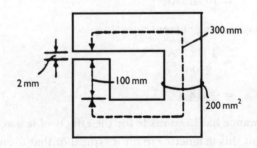

Fig. 51

In the gap $\qquad \Phi = B \times A$

$$\frac{0.12}{10^3} = \frac{B \times 200}{10^6} \quad \text{(Note conversions of milliweber to weber and mm}^2 \text{ to m}^2\text{)}$$

$$B = \frac{0.12 \times 10^6}{200 \times 10^3}$$

$$= 0.6\,\text{T}$$

and $\qquad \dfrac{B}{H} = \mu_0$

$$\frac{0.6}{H} = 4\pi \times 10^{-7}$$

$$H = \frac{0.6}{4\pi \times 10^{-7}}$$

m.m.f. $= H \times l$

$$= \frac{0.6}{4\pi \times 10^{-7}} \times \frac{2}{1000}$$

$$= 954.8\,\text{At}$$

In the absence of leakage and fringing the flux density in the steel is the same as that in the air gap, that is 0.6 T. By projecting the BH curve backwards slightly the necessary magnetizing force is found to be approximately 130 At/m.

The total length of steel is $300 + 100 = 400$ mm

The m.m.f. required for the steel is

$$130\,\frac{\text{At}}{\cancel{\text{m}}} \times \frac{400}{1000}\,\cancel{\text{m}} = 52\,\text{At}$$

The total m.m.f. $\qquad = 52 + 954.8$

$$= 1006.8\,\text{At}$$

$$\text{say } 1000\,\text{At}$$

The magnetizing current $= \dfrac{1000\,\text{A}\cancel{t}}{1000\,\cancel{t}}$

$$= \underline{1\,\text{A}}$$

1. Complete the following table:

Flux density B (T)		1.2	1.3				0.45
Cross-sectional area A (m^2)	0.5	0.006		0.65		0.002	0.035
Total flux Φ (Wb)					520 mWb	1000 mWb	

2. The air gap of a contactor is 25 mm in diameter. Calculate the total flux when the flux density is 0.9 T.

3. A magnetic circuit has a cross-sectional area of 0.75 m^2. Calculate the flux density when the total flux is 0.6 Wb.

4. The air gap of a moving-coil instrument is 15 by 25 mm. Determine the flux density when the total flux in the gap is 0.3 mWb.

5. Calculate the magnetizing force produced by a 350-turn coil carrying a current of 0.6 A when it is attached to a magnetic circuit 0.5 m long.

6. An air gap is 2 mm long and the magnetizing force required to set up a certain flux is 350 000 At/m. Find the number of ampere turns actually required for the gap.

7. The current available to set up a certain flux density in an air gap 5 mm long is 2 A. The magnetizing force required is found to be 200 000 At/m. Find the number of turns required on the coil.

8. A coil of 3500 turns is attached to a magnetic circuit 237 mm long. The magnetizing force required to set up a certain flux is 8000 At/m. Calculate the current which must flow in the coil.

9. Calculate the flux density produced in air by the following values of magnetizing force.
   (a) 20 000 At/m,
   (b) 105 000 At/m,
   (c) 750 000 At/m.

10. Plot a graph showing the relationship between flux density and magnetizing force for an air gap. Take values of H between 0 and 100 000 At/m.

11. The air gap in a certain magnetic circuit is 2.5 mm long and 350 mm^2 in cross-section. Calculate the magnetizing force required to produce a total flux of 0.1 mWb in the gap.

12. A certain magnetic circuit is energized by a coil of 250 turns. An air gap in the circuit is 25 mm in diameter and 1 mm long. Calculate the current which must flow in the coil to produce a total flux of 750 μWb in the gap. (Neglect the m.m.f. required for the steel.)

13. A circular steel core is 20 mm by 20 mm in cross-section and 0.15 m in mean diameter. It is provided with a coil of 500 turns. Using the figures given below calculate the current which must flow in the coil to produce a total flux of 0.5 mWb in the core.

Flux density B (T)	1.3	1.4	1.5
Magnetizing force H (At/m)	800	1250	2000

14. The magnetic circuit of a contactor is 0.3 m long and 15 mm by 25 mm in cross-section. The operating coil consists of 1500 turns of wire. Using the figures given below calculate the total flux produced in the core by a current of 0.067 A.

Flux density B (T)	0.2	0.8	1.0
Magnetizing force H (At/m)	100	240	400

15. If a radial air gap 1.5 mm wide is introduced into the core of question 14, calculate the current now required to produce the same total flux.

16. A steel ring of circular cross-section 150 mm^2 in area has a mean diameter of 85 mm. It is wound with 250 turns of wire. Assuming that the steel has relative permeability 500, calculate the current which must flow in the coil to produce a total flux in the steel of 0.035 mWb.

17. A coil of insulated wire of 400 turns and of resistance 0.25 ohm is wound tightly round an iron ring, and is connected to a d.c. supply of 4 volts. The iron ring is of uniform cross-sectional area 600 mm^2 and of mean diameter 0.15 m. The permeability of the ring may be taken as 450.

    Calculate the total flux in the iron.

    What would be the effect of an air gap in the ring?   (C & G)

18. The ring of question 17 now has a radial air gap 0.5 mm wide inserted in it. Calculate the current which is required to produce the same flux as before.

19. A coil of insulated wire of 500 turns and of resistance 4 ohms is closely wound on an iron ring. The ring has a uniform cross-sectional area of 700 mm^2 and a mean diameter of 0.25 m.

    Calculate the total flux in the ring when a d.c. supply at 6 V is applied to the ends of the winding. Assume a relative permeability of 550.

    Explain the general effect of making a small air gap by cutting the ring radially at one point.   (C & G)

20. The magnetic core of a contactor has the form shown in Fig. 52. It is made from 12 stampings each 1 mm thick. Using the BH values from question 18, determine the total m.m.f. required to produce a flux density of 0.9 T in the air gap.

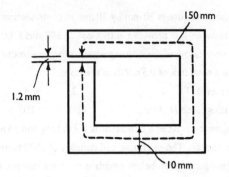

Fig. 52

21. A magnetic circuit has the form shown in Fig. 53. A winding on the centre limb has 1500 turns. It is required to produce a total flux in the air gap of 0.25 mWb. The relative permeability of the steel under these conditions is 12 000. Determine the magnetizing current.

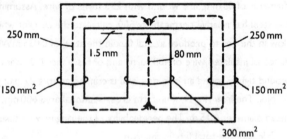

Fig. 53

Electromagnetism II

References: **Growth and decay of current in inductive circuits.**
Energy stored in magnetic circuits.
Discharge resistors.

DIRECT-CURRENT EXCITED CIRCUIT HAVING INDUCTANCE AND RESISTANCE IN SERIES

If such a circuit (Fig. 54) having inductance L henry in series with resistance R ohm is supplied at U volts d.c. the current at any instant t seconds after closing the switch S is found as follows:

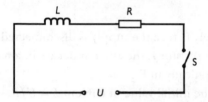

Fig. 54

(1) Calculate the final value of current $I = U/R$.
(2) Calculate the time constant $T = L/R$ (seconds).
(3) Draw graph axes to suitable scales as in Fig. 55. (The time required for the current to reach its maximum value may be taken as five times the time constant.)

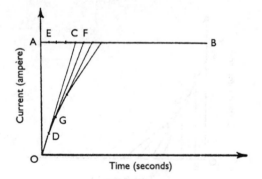

Fig. 55

(4) Draw the horizontal line AB so that OA $= I$. Mark off AC $= T$. Join OC. Select any point D on OC. Project upwards vertically from D to E on AB, mark EF $= T$. Join FD. Repeat for the new point G on DF and so on until the complete curve is traced.

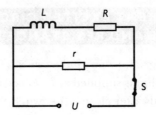

Fig. 56

If immediately before the supply is disconnected the coil is connected to a resistor r, the current decays in a manner illustrated graphically in Fig. 56.

(a) Calculate the initial value of current $I = U/R$.

(b) Calculate the time constant

$$T = \frac{L}{R+r}$$

(5) Draw axes to suitable scale (Fig. 57). Again the time required for the current to fall to zero may be taken as five times the time constant.

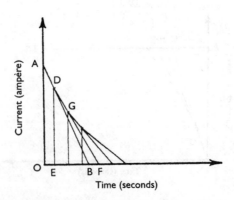

Fig. 57

(6) Mark $OA = I$ and $OB = T$. Join AB and select any point D on AB close to A. Project from D to E on the time axis and mark $EF = T$. Join DF. Select a new point G on DF and repeat the procedure until a complete curve is formed.

EXAMPLE 1 A relay has a coil of resistance 480 Ω and inductance 4.8 H. It is connected to a 240 V d.c. supply. At the end of 0.05 second the coil is short circuited and the supply disconnected.
(a) Draw curves showing the growth and decay of current plotted against time.
(b) If the relay closes when the current reaches 0.3 A increasing and opens when it reaches 0.15 A decreasing, estimate the total time for which the relay contacts are closed.

(a) Final value of current

$$I = \frac{U}{R}$$

$$= \frac{240}{480} = 0.5\,\text{A}$$

Time constant during increase of current

$$T = \frac{L}{R}$$

$$= \frac{4.8}{480}$$

$$= 0.01\,\text{second}$$

Since the coil is short-circuited during the decay period the total circuit resistance is the same and the time constant remains unchanged (Fig. 58).

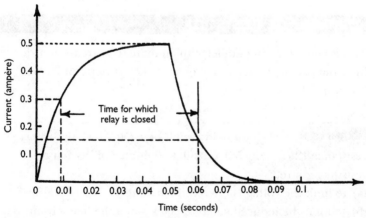

Fig. 58

(b) Total time for which relay is closed $= 0.05 - 0.0092 + 0.012$

$$= 0.0528 \text{ second}$$

$$= \underline{52.8 \text{ millisecond}}$$

ENERGY STORED

The energy stored in a circuit of inductance L henry when the current flowing is I ampere is

$$W = \tfrac{1}{2}LI^2 \text{ joule}$$

EXAMPLE 2 Calculate the maximum energy stored in the relay of the previous example.

The maximum current $I = \dfrac{U}{R}$

$$= \frac{240}{480} = 0.5\,\text{A}$$

$$W = \tfrac{1}{2}LI^2$$

$$= \tfrac{1}{2} \times 4.8 \times (0.5)^2$$

$$= \underline{0.6\,\text{J}}$$

DISCHARGE RESISTOR

This is connected in parallel with an inductive circuit in order to limit the rise in voltage which occurs when the circuit is interrupted.

EXAMPLE 3 The shunt field circuit of a 250 V d.c. motor has resistance $125\,\Omega$ (Fig. 59). Calculate the value of discharge resistor required:
(a) to limit the voltage between the field terminals to 500 V;
(b) to limit the induced e.m.f. to 500 V when the field circuit is broken.

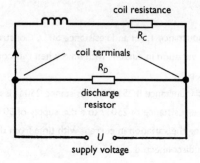

coil resistance
R_C

coil terminals
R_D

discharge
resistor

U
supply voltage

Fig. 59

When the supply is disconnected the current decays in the loop formed by R_C and R_D. At this instant the current in this loop is

$$I = \frac{U}{R_C}$$

In this case $I = \dfrac{250}{125}$

$$= \underline{2\,\text{A}}$$

(a) The voltage between the coil terminals is the voltage across the discharge resistor which is $I \times R_D$

i.e. $\quad 500 = I \times R_D$

$$R_D = \frac{500}{2}$$

$$= \underline{250\,\Omega}$$

(b) By applying Kirchhoff's law to the closed loop referred to above:

induced e.m.f. $\quad\quad E = IR_C + IR_D$

$$500 = 2 \times 125 + 2 \times R_D$$

$$2 \times R_D = 500 - 250$$

$$R_D = \frac{250}{2}$$

$$= \underline{125\,\Omega}$$

EXERCISE 9

1. A coil has inductance 1.5 H and resistance 50 Ω. Construct a curve showing the variation in current with time when this coil is connected to a 10 V d.c. supply.

2. A coil having inductance 0.25 H and resistance 250 Ω is connected in parallel with a resistance of 250 Ω to a d.c. supply of 50 V. Construct a curve showing the variation in current with time from the instant that the d.c. supply is disconnected.

3. The following figures show the variation in current with time when a coil of resistance 10 Ω and unknown inductance is connected to a d.c. supply of 100 V.

Current (A)	0	3.297	5.501	7.534	8.347	8.892	9.502
time (s)	0	0.4	0.8	1.4	1.8	2.2	3.0

   Given that the time constant L/R is equal to the time required for the current to rise to 0.632 of its maximum value, calculate the inductance of the coil.

4. A relay coil has resistance 1250 Ω and inductance 0.25 H. A d.c. voltage pulse which rises instantly to 50 V remains constant for 0.0015 second and then falls instantly to zero is applied to the coil. The voltage source may be assumed to have negligible resistance. Construct a curve showing the variation in current through the coil. If the relay closes when the current reaches 30 mA increasing and opens when the current falls to 15 mA decreasing, determine the total time for which the relay contacts are closed.

5. Calculate the maximum energy stored in each of the magnetic circuits of questions 12–15 inclusive in Exercise 8.*

6. The field system of a d.c. motor has four coils in series. Each coil has 1000 turns and resistance 20 Ω. When the supply voltage is 240 V the flux per pole is 0.05 Wb, when the voltage is 120 V the flux per pole is 0.03 Wb. Calculate the energy stored in the field system under normal running conditions when the voltage is 240 V.*

7. The field system of a 400 V d.c. motor has a total resistance 100 Ω. Calculate the value of a discharge resistor which will limit the induced e.m.f. to 1000 V when the field circuit is broken.

8. A 500 V d.c. generator has six poles connected in series. Each pole has 750 turns and resistance 15 Ω. A change in current of 2.5 A through the

winding produces a change in flux of 0.06 Wb. Calculate:

(a) the inductance of the field system;[*]

(b) the energy stored during normal working;

(c) the value of a discharge resistor necessary to limit the e.m.f. induced when the circuit is broken to 1000 V.

(d) the energy dissipated in the discharge resistor.

Direct current generator calculations

References: **Generator e.m.f. equation.**
Terminal voltage/load current.
Characteristics.

The e.m.f. generated by a d.c. generator is

$$E = \frac{\Phi \times Z \times n \times p}{c}$$

E is the e.m.f. in volts

Φ is the useful flux per pole in weber

Z is the number of armature conductors

p is the number of poles

c is the number of armature circuits

$c = p$ for a lap winding

$c = 2$ for a wave winding

n is the speed in rev/s

[*] Hint: Use the definition $\quad \text{Inductance} = \dfrac{\text{change in flux linkages}}{\text{change in magnetizing current}}$

The terminal voltage is given by

$$U = E - I_a R_a$$

U is the terminal voltage in volts
E is the generated e.m.f. in volts
I_a is the armature current in ampères
R_a is armature resistance in ohms

EXAMPLE 1 Calculate the terminal voltage of a d.c. generator
to which the following particulars refer, when it carries a load
of 25 A.

Field: four poles separately excited, useful flux per pole:
            0.025 Wb

Armature: 800 conductors, lap connected, total resistance
            0.1 Ω, speed of rotation 15 rev/s

$$E = \frac{\Phi \times Z \times n \times p}{c}$$

$$= \frac{0.025 \times 800 \times 15 \times 4}{4}$$

$$= 300 \,\text{V}$$

$$U = E - I_a R_a$$

$$= 300 - 25 \times 0.1$$

$$= 300 - 2.5$$

$$= \underline{297.5 \,\text{V}}$$

EXAMPLE 2 A d.c. generator delivers a load current of 50 A at a
terminal voltage of 200 V. The total resistance of its armature
circuit is 0.15 Ω and there is a 2 volt drop at the brushes.
Calculate:

(a) the generated e.m.f.;

(b) the speed at which it must be driven, given the following
information:

Number of poles 6, separately excited.

Useful flux per pole 0.04 Wb.

Number of armature conductors 600, lap connection.

(a)
$$U = E - I_a R_a$$

$$200 = E - 50 \times 0.15 - 2$$

$$E = 200 + 2 + 7.5$$

$$= \underline{209.5\,\text{V}}$$

(b)
$$E = \frac{\Phi \times n \times Z \times p}{c}$$

$$209.5 = \frac{0.04 \times n \times 600 \times 6}{6}$$

$$= 24n$$

$$n = \frac{209.5}{24}$$

$$= \underline{8.73\,\text{rev/s}}$$

Many syllabuses do not require the full use of the e.m.f. equation, so the following examples are based upon typical examination questions.

EXAMPLE 3 A load of 20.2 kW at 230 V is supplied by a shunt wound d.c. generator. The shunt field has a resistance of 110 Ω and the armature a resistance (including brushes) of 0.25 Ω. Brush contact volts drop is 1.9 V.
Calculate the:
(a) armature current
(b) generated e.m.f.
(c) total electrical losses.

(a) Load current $(I) = \dfrac{P}{U} = \dfrac{20.2 \times 1000}{230}$

$$= 87.83\,\text{A}$$

Field current $(I_f) = \dfrac{U}{R_f} = \dfrac{230}{110}$

Armature current $(I_a) = I + I_f = 87.83 + 2.09$

$$= \underline{89.92\,\text{A}}$$

(b) Terminal voltage = generated e.m.f.

                      − armature volts drop

                      − brush contact volts drop.

Thus $U = E - I_a R_a - 1.9$

and $E = U + I_a R_a + 1.9$

$$= 230 + (89.92 \times 0.25) + 1.9$$

$$= 230 + 22.48 + 1.9$$

$$= \underline{254.38\,\text{V}}$$

(c) Copper losses in armature winding $I_a^2 R_a = 89.92^2 \times 0.25$

$$= 2021.4\,\text{W}$$

Copper losses in field winding $(I_f^2 R_f)\quad = 2.09^2 \times 110$

$$= 480.49\,\text{W}$$

Total electrical losses $\qquad\qquad = 2012.4 + 480.49$

$$= 2501.89\,\text{W}$$

EXAMPLE 4 A 200 V d.c. generator runs at 25 rev/s and supplies a load current of 15 A. If the input torque to the generator shaft is 26 Nm, determine the generator:

(a) efficiency at this load

(b) power loss at this load

(a) efficiency $= \dfrac{UI}{T(2\pi n)} \times 100$

$$= \dfrac{200 \times 15 \times 100}{26 \times 2\pi \times 25}$$

$$= 73.46\%$$

(b) input power = output power + losses

    thus $T(2\pi n) = UI + \text{losses}$

    so losses $= T(2\pi n) - UI$

$$= (26 \times 2\pi \times 25) - (200 \times 15)$$

$$= 4084 - 3000$$

$$= 1084\,\text{W}$$

1. Explain the function of a commutator in a direct-current generator.

    A load of 19.2 kW is supplied from the terminals of a two-pole d.c. shunt generator at 240 V. The shunt winding of the generator has a resistance of 96 Ω, and the resistance of the armature is 0.2 Ω. There is a brush-contact volts drop of 2 V.

    Calculate:

    (a) the current in the armature,

    (b) the e.m.f. generated, and

    (c) the copper losses in the machine. (C & G)

2. A d.c. generator supplies a load of 80 A through cables which have total resistance 0.05 Ω. Its armature circuit resistance is 0.1 Ω and there is a 1 V drop at the brushes. Calculate its generated e.m.f. in order that the voltage at the load end shall be 200 V.

3. A d.c. generator to which the following particulars refer delivers 40 A at a terminal voltage of 500 V. Calculate the useful flux per pole.

    Number of poles: 4;

    Armature winding: 700 conductors, wave connected total resistance 0.1 Ω;

    Speed of rotation: 10 rev/s.

4. Calculate the speed at which a d.c. generator must be driven in order to deliver 30 A at a terminal voltage of 450 V given the following details:

    Number of poles: 4;

    Number of armature conductors: 600;

    Useful flux per pole: 0.03 Wb;

    Type of winding: wave;

    Resistance of armature: 0.15 Ω.

5. The following particulars refer to a certain shunt-connected d.c. generator:

    Number of poles: 6;

    Armature winding: 750 conductors lap-connected, total resistance 0.5 Ω;

    Useful flux per pole: 0.03 Wb;

    Resistance of field winding: 150 Ω;

    Brush-contact voltage drop: 2 V.

    Calculate the speed at which the machine must be driven in order to generate 12.5 kW at a terminal voltage of 250 V.

6. The open-circuit voltage of a d.c. generator is 250 V. Calculate its open-circuit voltage when the speed is increased by 20% and the flux is reduced by 15%.

7. A d.c. generator delivers 20 A at 250 V. Its efficiency is 72%. Calculate (a) the power required to drive it, (b) the torque required in Nm if its speed is 16 rev/s.

8. A d.c. shunt generator has a shunt field resistance of 120 Ω and an armature resistance of 0.1 Ω, when supplying a load of 28 A at a terminal voltage of 240 V. Calculate the:

   (a) shunt field circuit current;

   (b) armature current;

   (c) e.m.f. generated.

9. A 200 V d.c. generator running at 25 rev/s supplies a current of 15 A. If the input torque to the generator shaft is 26 Nm, determine the:

   (a) efficiency of the generator;

   (b) generator power loss.

10. A shunt-wound d.c. generator supplies a load of 18.4 kW at 230 V through a two-core cable of total resistance 0.05 Ω. The resistance of the armature is 0.03 Ω and there is a total brush contact drop of 2 V. The resistance of the field winding is 78 Ω. Calculate the generator terminal voltage and the generated e.m.f.   (C & G)

11. A 240 V generator supplies a load of 25 kW. The shunt field resistance is 160 Ω and the armature resistance is 0.12 Ω. Assuming a brush contact voltage drop of 2 V, find the:

    (a) armature current;

    (b) generated e.m.f.;

    (c) copper losses in the windings.   (C & G)

12. Tests being conducted on a diesel engine driven shunt-wound d.c. generator produce the following results;

    (a) input torque to generator = 23 Nm

    (b) output voltage = 150 V

    (c) armature current = 14 A.

    The generator shunt field regulator is now set to reduce the flux to 80% and the input torque now increases to 33 Nm. What will the armature current be at this new torque.

13. A shunt-wound d.c. generator supplies a load of 18 kW at 240 V through a two-core cable of total resistance 0.055 Ω. The resistance of the

armature is 0.033 Ω and there is a total voltage drop 2.4 V in the brush gear. The resistance of the shunt field is 79 Ω. Calculate the:

(a) load current;

(b) voltage required at the generator terminals;

(c) generator field current;

(d) generator armature current;

(e) generated e.m.f.

Direct current motor calculations

References: **Direct current motor, total torque**
and shaft torque.
Factors influencing speed.
Efficiency.
Starting resistance.

The total torque of a d.c. motor is given by

$$T = \Phi \times I_a \times \frac{Zp}{2\pi c} \text{ Nm}$$

where Φ is the useful flux per pole (Wb)

I_a is the armature current (A)

Z is the number of armature conductors

p is the number of poles

c is the number of armature circuits

$(c = p$ for lap windings

$c = 2$ for wave windings).

EXAMPLE 1 A d.c. motor develops a total torque 150 Nm when its armature current is 25 A and the useful flux per pole is 0.25 Wb. Calculate the total torque when the armature current increases to 35 A and the flux is reduced to 0.2 Wb.

We may replace $Zp/2\pi c$ by k, a constant, in the torque equation.

Then $T = k \times \Phi \times I_a$

If T_1 is the torque corresponding to flux Φ_1 and current I_1 and T_2 is the torque corresponding to flux Φ_2 and current I_2

$$T_1 = k \times \Phi_1 \times I_1 \tag{i}$$

and $\quad T_2 = k \times \Phi_2 \times I_2 \tag{ii}$

Dividing (ii) by (i)

$$\frac{T_2}{T_1} = \frac{\Phi_2 \times I_2}{\Phi_1 \times I_1} \quad k \text{ cancelling}$$

Calling T_2 the new value of torque

$$\frac{T_2}{150} = \frac{0.2 \times 35}{0.25 \times 25}$$

$$T_2 = \frac{150 \times 0.2 \times 35}{0.25 \times 25}$$

$$= \underline{168 \, \text{Nm}}$$

EXAMPLE 2 Details of a d.c. motor are as follows:

Number of poles: 6

Number of armature conductors: 700

Useful flux per pole: 0.056 Wb

Type of armature winding: lap.

Its armature current is 12 A when it delivers 4.103 kW to a machine at 11 rev/s. Calculate the total torque and the shaft torque. Express these in the same units.

Total torque is calculated from the formula

$$T = \Phi \times I_a \times \frac{Zp}{2\pi c}$$

$$= 0.056 \times 12 \times \frac{700 \times 6}{2\pi \times 6}$$

$$= \underline{74.85\,\text{Nm}}$$

Shaft torque is calculated from the formula

$$P = 2\pi n T$$

where P is in watts

n is the speed in rev/s

T is the torque in Nm

$$4103 = 2\pi \times 11 \times T$$

$$T = \frac{4103}{\pi \times 11}$$

$$= \underline{59.36\,\text{Nm}}$$

BACK E.M.F. AND SPEED

When the motor runs on load its speed is such that the equation $U = E + I_a R_a$ is satisfied.

U is the armature supply voltage

E is the back e.m.f.

I_a is the armature current

R_a is the resistance of the armature circuit

and E is given by the generator e.m.f. formula

$$E = \Phi \times n \times Z \times p$$

which is simplified to

$$E = k \times \Phi \times n \quad \text{where } k \text{ is a constant.}$$

Many syllabuses do not require the full use of the torque equation, so the following examples are based upon typical examination questions.

EXAMPLE 3 A 2.5 kW 250 V d.c. shunt-wound motor runs at 16 rev/s and takes a current of 14 A when developing full-load output. The armature resistance is 0.4 Ω and the field resistance is 160 Ω.

Calculate at full load the:

(a) armature current;
(b) back e.m.f.;
(c) efficiency;
(d) shaft torque.

(a) Armature current (I_a) = supply current − field current (I_f)

Now $I_f = \dfrac{U}{R_f}$ where R_f is the resistance of the shunt field

$$I_f = \frac{250}{160} = 1.56 \, \text{A}$$

and $I_a = 14.0 - 1.56 = 12.44 \, \text{A}$

(b) Back e.m.f. is $E_b = U - I_a R_a$, where R_a is the resistance of the armature

Then $E_b = 250 - (12.44 \times 0.4)$

$\qquad = 250 - 4.98$

$\qquad = 245.02 \, \text{V}$

(c) Efficiency $= \dfrac{\text{output}}{\text{input}} = \dfrac{2500}{250 \times 14}$

$\qquad = \dfrac{2500 \times 100}{3500}$

$\qquad = 71.4\%$

(d) Shaft torque $= \dfrac{2500}{2\pi \times 16}$

$\qquad = 24.87 \, \text{Nm}$

EXAMPLE 4 A 20 kW, 240 V d.c. shunt motor has a full-load efficiency of 80%. The shunt field resistance is 106 Ω and the armature resistance (including brushes) is 0.12 Ω. Assuming a brush contact voltage drop of 2 V, determine at full load the:

(a) armature current;

(b) generated back e.m.f.;

(c) total copper losses. (C & G)

(a) Input to motor $= \dfrac{20 \times 100}{80}$

$= 25\,\text{kW}$

Full load current $= \dfrac{25\,000}{240}$

$= 104.17\,\text{A}$

Field current $(I_f) = \dfrac{240}{160}$

$= 1.5\,\text{A}$

Armature current $(I_a) = 104.17 - 1.5$

$= \underline{102.67\,\text{A}}$

(b) Back e.m.f. $E_b = U - I_a R_a - \text{brush volts drop}$

$= 240 - (102.67 \times 0.12) - 2$

$= 240 - 12.32 - 2$

$= \underline{225.68\,\text{V}}$

(c) Total copper losses $= \text{Cu loss in field}$
$+ \text{Cu loss in armature}$

$= I_f^2 R_f + I_a^2 R_a$

$= (1.5^2 \times 160) + (102.67^2 \times 0.12)$

$= 360 + 1265$

$= \underline{1.625\,\text{kW}}$

EXAMPLE 5 A d.c. shunt motor-generator set running at 12 rev/s supplies a current of 13 A to an external load connected to the generator terminals. At this load the motor, connected to a 90 V supply, takes a current of 12 A.

Details of the machines at this load are as follows:

	Motor	Generator
armature resistance	$0.4\,\Omega$	$0.2\,\Omega$
field current	$2\,\text{A}$	$1\,\text{A}$
constant losses	$114\,\text{W}$	$46\,\text{W}$

Calculate the:
(i)   mechanical power delivered to the motor shaft;
(ii)   motor output torque;
(iii) efficiency of the motor-generator set.   (C & G)

(i)   Motor.

Field copper losses (I^2R_f) $= 90 \times 2$ $= 180\,\text{W}$

Armature current $= 12 - 2$ $= 10\,\text{A}$

Armature copper losses $(I^2R_a) = 10^2 \times 0.4 = 40\,\text{W}$

Electrical input (W) $= 90 \times 12$ $= 1080\,\text{W}$

$\therefore$ Output power from motor $= 1080 - (114 + 180 + 40)$

$= \underline{746\,\text{W}}$

(ii) Motor output torque $= \dfrac{746}{2\,\Omega \times 12} = \underline{9.9\,\text{Nm}}$

(iii) Generator.
Mechanical input to
  generator $= 746\,\text{W}$

Armature current $= 13 + 1$ $= 14\,\text{A}$

Armature copper losses $(I^2R_a) = 14^2 \times 0.2$ $= 39.2\,\text{W}$

Armature output $= 746 - (46 + 39.2)$

$= 660.8\,\text{W}$

Generator output voltage $= \dfrac{660.8}{14}$ $= 47.2\,\text{V}$

Generator output power $= 13 \times 47.2$ $= 613.6\,\text{W}$

$\therefore$ Efficiency of motor-generator $= \dfrac{613.6}{1080} \times 100 = \underline{56.8\%}$

EXAMPLE 6 On no load the speed of a d.c. motor is 16 rev/s. Calculate the speed when the load is such that the armature current is 25 A. The terminal voltage is constant at 400 V, the armature resistance is 0.15 Ω and armature reaction is to be neglected.

If the no-load armature current is neglected the equation

$$U = E + I_a R_a$$

reduces to

$$U = E$$

i.e. the back e.m.f. is equal to the supply voltage.

Also if armature reaction is neglected the flux is constant and

$$E = kn$$

as shown previously, if E_1, n_1: E_2, n_2 are corresponding values

$$\frac{E_2}{E_1} = \frac{n_2}{n_1}$$

Thus if $n_1 = 16$ and n_2 is to be calculated, first calculate E_2

$$U = E_2 + I_a R_a$$

$$400 = E_2 + 25 \times 0.15$$

$$E_2 = 400 - 3.75$$

$$= 396.25 \text{ V}$$

Then $\dfrac{n_2}{16} = \dfrac{396.25}{400}$ $(E_1 = U = 400)$

$$n_2 = 16 \times \frac{396.25}{400}$$

$$= \underline{15.85 \text{ rev/s}}$$

EFFICIENCY TESTS: DIRECT METHOD

EXAMPLE 7 A test on a d.c. motor yielded the following results:
Electrical input
   Terminal voltage 460 V
   Supply current 18.9 A.

Mechanical output from brake test

   Diameter of brake pulley 0.4 m

   Nett brake load 320 N

   Speed 16 rev/s.

Calculate the efficiency at this condition

Torque $\qquad T = F \times r$

where F is the nett brake load (N)

      r is the pulley radius (m)

$$T = 320 \times \frac{0.4}{2}$$

$$= 62 \, \text{Nm}$$

Mechanical output $= 2\pi n T \, \text{W}$

$$= 2\pi \times 16 \times 62$$

$$= 6233 \, \text{W}$$

Electrical input $\quad = 460 \times 18.9$

$$= 8694 \, \text{W}$$

Efficiency $\qquad = \dfrac{\text{output}}{\text{input}}$

$$= \frac{6233}{8694}$$

$$= 0.7169$$

$$\text{or } \underline{71.69\%}$$

INDIRECT METHOD

EXAMPLE 8 A no-load test on a d.c. shunt motor produced the following results:

Supply voltage	400 V
Armature current	3.5 A
Speed	12.5 rev/s
Armature resistance	0.8 Ω
Field resistance	120 Ω.

Predict the efficiency and speed of the machine when the load is such that the armature current is 40 A.

This is the 'summation of losses' method of determining efficiency. The various losses to be calculated are:

(a) field circuit copper loss $\Big\}$ assumed constant
(b) iron friction and windage loss

(c) armature copper loss varying with the square of the load.

For (a) field copper loss $= \dfrac{U^2}{R_f}$

$$= \frac{400^2}{120}$$

$$= 1333 \, \text{W}$$

(b) This is the total input to the armature on no load:

$$= 400 \times 3.5$$

$$= 1400 \, \text{W}$$

Total constant loss $= 1333 + 1400$

$$= 2733 \, \text{W}$$

For (c) corresponding to an armature current of 40 A the armature copper loss

$$= I_a^2 R_a$$

$$= 40^2 \times 0.8$$

$$= 1280 \, \text{W}$$

Total losses $= 2733 + 1280$

$$= 4013 \, \text{W}$$

Efficiency $= \dfrac{\text{input}}{\text{input} + \text{losses}}$

$$= \frac{400 \times 40}{(400 \times 40) + 4013}$$

$$= 0.795 \text{ or } \underline{79.5\%}$$

The speed is calculated as in Example 4.

Thus on no load

$$U = E_1 \text{ (approximately)}$$

or $E_1 = 400$

and on load of 40 A,

$$400 = E_2 + 40 \times 0.8$$

$$E_2 = 400 - 32$$

$$= 368\,\text{V}$$

and $\dfrac{n_2}{n_1} = \dfrac{E_2}{E_1}$

where n_2 is the required speed and n_1 is the no-load speed

$$n_2 = 12.5 \times \frac{368}{400}$$

$$= \underline{11.5\,\text{rev/s}}$$

(What assumptions have been made in this example?)

STARTING RESISTANCE

EXAMPLE 9   The armature of a 400 V, 30 kW motor has resistance 0.65 Ω. Assuming that the full-load efficiency of the motor is 75%, calculate the value of the series resistor which will limit the starting current to $1\frac{1}{2}$ times its full-load value.

The full-load current is calculated as follows:

Full-load output (watts) $= 30 \times 1000$

Full-load input $= 30 \times 1000 \times \dfrac{100}{75}\,\text{W}$

Full-load current $= \dfrac{30 \times 1000 \times 100}{400 \times 75}$

$= 100\,\text{A}$

$1\frac{1}{2} \times$ full-load current $= \frac{3}{2} \times 100$

$= 150\,\text{A}$

In the absence of back e.m.f.

$$U = I_a R_a$$

where R_a is the total armature circuit resistance

$$400 = 150 \times R_a$$

$$R_a = \frac{400}{150}$$

$$= 2.667\,\Omega$$

Value of the additional series resistance is thus

$$2.667 - 0.65 = 2.017\,\Omega$$

$$\text{or } \underline{2.02\,\Omega}$$

EXERCISE 11

1. A d.c. motor develops a torque of 200 Nm under given conditions of flux and armature current. Calculate the torque when the armature current falls by 12% and the flux increases by 8%.

2. A d.c. shunt motor connected to a 240 V supply has a no-load speed of 24.6 rev/s. The current input at no load is 5 A, and at full load 42 A. The armature resistance is 0.2 Ω, and the shunt winding resistance is 160 Ω.

   Calculate the speed of the motor at full load. (C & G)

3. The armature resistance of a d.c. motor is 0.1 Ω. On no load it runs at a speed of 20 rev/s from a 250 V supply. Calculate the speed at which it will run when its armature current is 15 A. The supply voltage and magnetic flux remain constant.

4. Calculate the total torque developed by a d.c. motor to which the following details refer:

   Number of poles: 4;

   Number of armature conductors: 740;

   Useful flux per pole: 0.3 Wb;

   Type of winding: wave;

   Armature current: 20 A.

5. Calculate the value of armature current required in the motor of question 4 in order for it to develop a total torque of 1200 Nm, all other conditions remaining constant.

6. A d.c. motor runs at 13 rev/s when its armature current is 25 A and the terminal voltage is 250 V. Its armature resistance is 0.12 Ω. Calculate the speed at which it must be driven as a generator in order for it to deliver 25 A at a terminal voltage of 250 V, the flux remaining constant.

7. A 460 V, d.c. shunt motor running on no load at 46.6 rev/s, takes a current of 6 A. The resistance of the field winding is 230 Ω and the resistance of the armature circuit is 0.3 Ω.

    Calculate the speed of the motor when it runs with a full-load input of 35 A. Assume that the field current remains constant.

8. A 250 V d.c. shunt motor runs at 11 rev/s when the input current is 42 A. The armature circuit resistance is 0.25 Ω, the shunt field resistance is 62.5 Ω and the sum of the iron, friction and windage losses is 1400 W.

    Calculate the:

    (a) motor output,

    (b) motor efficiency,

    (c) shaft torque.   (C & G)

9. A 2.5 kW, 220 V d.c. shunt-wound motor runs at 18 rev/s and takes a current of 18 A when developing full-load output. The armature resistance is 0.3 Ω and the field resistance is 180 Ω.

    Calculate at full load the:

    (a) armature current,

    (b) back e.m.f.,

    (c) efficiency,

    (d) torque.

10. The resistance of armature and shunt field respectively of a d.c. motor are 0.15 Ω and 250 Ω. On no load when the armature current may be neglected, the speed is 10 rev/s. Calculate the speed when the load is such that the total input current is 35 A. Assume constant terminal voltage of 400 V.

11. A d.c. shunt motor connected to a 240 V d.c. supply has a no-load speed of 24.5 rev/s. The current input at no load is 4 A, and at full load the current input is 38 A. The shunt field-windings have a resistance of 150 Ω and the resistance of the armature is 0.25 Ω.

    Calculate the speed at full load.

12. The following readings were taken during a brake test on a d.c. motor.

    Brake load   196 N at an effective radius of 460 mm
    Speed   20 rev/s
    Electrical input   57 A at 250 V

    Calculate the efficiency of the motor at this load.

13. The following readings were taken during a brake test on a d.c. motor.

    Speed of motor   24 rev/s
    Effective diameter of brake pulley   250 mm
    Effective pull at circumference of pulley   220 N
    Electrical input   5800 W

    Find the power output and the efficiency at this load.   (C & G)

14. A d.c. motor runs at 12.5 rev/s on no load when its terminal voltage is 250 V. The input is 800 W. The armature resistance is 0.6 Ω. Calculate the efficiency under the same conditions of terminal voltage and speed when the armature current is (i) 30 A, (ii) 50 A.

15. The armature of a d.c. motor has a resistance of 0.45 Ω and takes a full-load current of 50 A from a 400 V supply. Calculate the value of additional series resistance required to limit the starting current to $1\frac{1}{2}$ times the full-load value.

16. A 200 V, 7.5 kW d.c. motor operates with a full-load efficiency of 70%. Its armature resistance is 0.8 Ω. Calculate the value of series resistance required to limit the starting current to $1\frac{1}{2}$ times the full-load value.

17. Calculate the ohmic value of a starting resistor for the following d.c. shunt motor:

    Output   14 920 W
    Supply   240 V
    Armature resistance   0.25 Ω
    Efficiency at full load 86%

    The starting current is to be limited to $1\frac{1}{2}$ times the full-load current. Ignore the current in the shunt winding.

18. A 460 V d.c. motor runs at 15.8 rev/s and develops 7460 W with an efficiency of 85%. The armature resistance is 0.2 Ω. Calculate the value of resistor which when connected in series with the armature will reduce the speed to 11.7 rev/s, the armature current remaining constant.

Utilization of electric power I

References: **Torque, work and power**.
  Efficiency of machines.
  Power in d.c. and a.c. circuits.

EXAMPLE 1 A load of 100 kg is raised through a vertical
distance of 12 m in 15 s by a hoist. The efficiency of the hoist
gearing is 30% and that of the driving motor is 80%. Calculate
the electrical power input to the motor.

The force required to lift a load of 1 kg against the effect of
gravity is 9.81 N. (This is the same as saying that 1 kgf is
equivalent to 9.81 N.)

Work done $W =$ distance × force

$$= 12\,\text{m} \times 100 \times 9.81\,\text{N}$$

$$= 11\,772\,\text{mN}$$

and 1 mN (or 1 Nm) $= 1\,\text{J}$ (joule)

Thus the work done on the load

$$= 11\,772\,\text{J}$$

and the work done per second

$$= \frac{11\,772}{15}\,\text{J/s}$$

$$= 784.8\,\text{W} \quad (\text{for } 1\,\text{J/s} = 1\,\text{W})$$

The input to the hoist gear which is the driving motor output

$$= 784.8 \times \frac{100}{30}$$

$$= 2616\,\text{W}$$

The input to the driving motor

$$= 2616 \times \frac{100}{80}$$

$$= 3270 \, \text{W}$$

$$\text{or } \underline{3.27 \, \text{kW}}$$

EXAMPLE 2 A hoist raises a mass of 3000 kg a distance of 20 m in 50 s. It is driven by a 400 V 50 Hz three-phase 8-pole cage rotor induction motor, working with a power factor of 0.8. The efficiencies of the motor and hoist are 90% and 65% respectively. (Take g as 9.81 m/s².)

(a) Calculate the:
- (i) output of the hoist in kW,
- (ii) output of the motor in kW,
- (iii) input to the motor in kVA,
- (iv) line current.

(b) If the drum of the winding gear has an average diameter of 0.1 m find the gear ratio required between the motor and the drum. Assume the slip of the motor to be 8%.

(a)(i) Output of the hoist $= \dfrac{3000 \times 9.81 \times 20}{50}$

$$= 11.77 \, \text{kW}$$

(ii) Output of motor $= \dfrac{11.77}{0.65}$

$$= 18.11 \, \text{kW}$$

(iii) Input to motor $= \dfrac{18.11}{0.9}$

$$= 20.12 \, \text{kW}$$

Input in kVA $= \dfrac{20.12}{0.8}$

$$= 25.15 \, \text{kVA}$$

(iv)   Line current $= \dfrac{25.15 \times 10^3}{\sqrt{3} \times 400}$

$= 36.3\,\text{A}$

(b) Speed of winding drum $= \dfrac{20}{3.14 \times 0.1 \times 50}$

$= 1.27\,\text{rev/s}$

Speed of motor $= \dfrac{50 \times 92}{4 \times 100}$

$= 11.5\,\text{rev/s}$

So gear ratio necessary $= \dfrac{11.5}{1.27}$

$= 9.05 : 1$

EXAMPLE 3   A pump raises $0.0075\,\text{m}^3$ of water per second through a vertical height of 30 m. Its efficiency is 70% and it is driven by a 400 V three-phase motor of efficiency 85% and power factor 0.8. Calculate the line current to the motor.

$1\,\text{m}^3$ of water weighs $1000\,\text{kgf}$

Following the previous example:

Work done by the pump per second

$= \dfrac{30 \times 0.0075 \times 1000 \times 9.81}{1}$

$= 2207\,\text{Nm/s}$ or   J/s   or   W

Electrical input to driving motor

$= 2207 \times \dfrac{100}{70} \times \dfrac{100}{85}$

$= 3709\,\text{W}$

The power in a three-phase circuit

$P = \sqrt{3}\,U_L I_L \cos\phi$

$3709 = \sqrt{3} \times 400 \times I_L \times 0.8$

$$I_L = \frac{3709}{\sqrt{3} \times 400 \times 0.8}$$

$$= \underline{6.69\,\text{A}}$$

EXAMPLE 4 The cutting tool of a lathe experiences a force of 400 N when a workpiece 150 mm in diameter is rotated against it at a speed of 2.66 rev/s (Fig. 60). Calculate the power absorbed by the cutting operation.

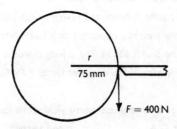

Fig. 60

$$\text{Torque} = \text{force} \times \text{radius of workpiece}$$

$$= 400 \times \frac{150}{2 \times 1000}$$

$$= 30\,\text{Nm}$$

Since

$$P = 2\pi nT\,\text{W}$$

where T is the torque in Nm and n the speed in rev/min

$$P = 2\pi \times 2.66 \times 30$$

$$= \underline{501.4\,\text{W}}$$

EXERCISE 12

1. A conveyor raises 250 kg of goods per minute through a vertical distance of 25 m. It is driven through a gearbox of efficiency 62%. Determine the power output of the driving motor.

2. A pump rotating at 16 rev/s raises 0.015 m^3 of water per second through a vertical distance of 50 m. Calculate the torque required to drive it.

3. Calculate the torque required if the pump of question 2 is used to pump oil of relative density 0.78 under the same conditions.

4. A generator of efficiency 82% delivers 75 A at 120 V. Calculate the torque required to turn it at 16 rev/s.

5. A crane raises a load of 2000 kg at a speed of 0.45 m/s. The efficiency of the gearing is 55% and it is driven by a 250 V d.c. motor of efficiency 72%. Calculate the armature current.

6. The chuck of a lathe is driven at 2 rev/s through gearing which is 60% efficient. During a turning operation on a workpiece 75 mm in radius the pressure on the tool is 650 N. The driving motor is a three-phase 400 V machine of efficiency 85% and power factor 0.75. Calculate the line current.

7. A brake test on a three-phase motor yielded the following results:

line current 12 A	speed 16 rev/s
line voltage 400 V	brake tension tight side 227 N;
wattmeter readings 4986 W,	slack side 40 N
982 W	brake pulley diameter 0.5 m.

Calculate the efficiency and the power factor at this load.

8. The cage of a lift weighs 1000 kg and can carry a maximum load of 1500 kg. The hoisting ropes pass over a driving sheave 1 m in diameter and are attached to a balance weight equal to the weight of the cage plus 40% of the maximum load. Assuming that the cage is three-quarters loaded and travelling upwards at a speed of 0.66 m/s, calculate:

(a) the speed at which the driving sheave rotates;

(b) the torque required to turn it.

9. If the sheave of question 8 is driven by a 460 V d.c. motor of efficiency 70% through gearing of efficiency 65%, calculate the power output of the motor and the current taken by it.

10. A d.c. generator delivers 85 A at 120 V and is 78% efficient. It is driven at 16 rev/s by a three-phase 415 V motor of efficiency 82% and power factor (corrected) of 0.9. Calculate the line current to the motor and its shaft torque.

Transformers, etc.

References: **E.m.f. equation.**   **Ratio and proportion.**
Phasor diagrams.   **Phasor resolution and**
No-load current.   **combination.**
Efficiency.   **Open- and short-**
System short-circuit   **circuit tests.**
conditions.   **Regulation.**

E.M.F. EQUATION

The e.m.f. of a transformer is given by

$$E = 4.44 \times f \times \Phi_{\max} \times N$$

where f is the frequency (Hz)

   $\Phi_{\max}$ is the maximum value of core flux (Wb)

   E is the e.m.f. induced in the winding having
   N turns (V).

Thus for the primary

$$E_P = 4.44 \times f \times \Phi_{\max} \times N_P$$

and for the secondary

$$E_S = 4.44 \times f \times \Phi_{\max} \times N_S$$

where E_P, N_P, E_S, N_S, refer to primary and secondary
respectively.

   In each case the e.m.f. and the terminal voltage are equal if the
winding resistance and reactance are ignored.

EXAMPLE I   Calculate the maximum value of flux in the core of
a transformer having 2000 primary turns and supplied at 230 V,
50 Hz.

$$E_P = 4.44 \times f \times \Phi_{\text{max}} \times N_P$$

$$230 = 4.44 \times 50 \times \Phi_{\text{max}} \times 2000$$

$$\Phi_{\text{max}} = \frac{230}{4.44 \times 50 \times 2000}$$

$$= \underline{0.000\,518\,\text{Wb}}\ (0.518\,\text{mWb})$$

EXAMPLE 2 If the maximum flux density in the core of the transformer of Example 1 is not to exceed 0.5 T, calculate the cross-sectional area of the core. Draw the no-load phasor diagram.

$$\Phi_{\text{max}} = B_{\text{max}} \times A$$

where B_{max} is the maximum flux density and A is the cross-sectional area of the core (m^2)

$$0.000\,518 = 0.5 \times A$$

$$A = \frac{0.000\,518}{0.5}$$

$$= \underline{0.00104\,\text{m}^2}$$

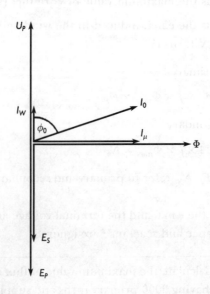

Fig. 61 No-load phasor diagram

Φ is the flux

E_P and E_S are primary and secondary induced e.m.f.s respectively

U_P is the primary applied voltage

I_0 is the no-load current

I_W and I_μ are its active and reactive components respectively

$\cos \phi_0$ is the no-load (or open-circuit) power factor.

EXAMPLE 3 The no-load current of a 250/50 V transformer is 3 A at 0.2 p.f. lagging. Draw the no-load phasor diagram accurately to scale and determine the active and reactive components of the no-load current.

The phasor diagram is shown in Fig. 62.

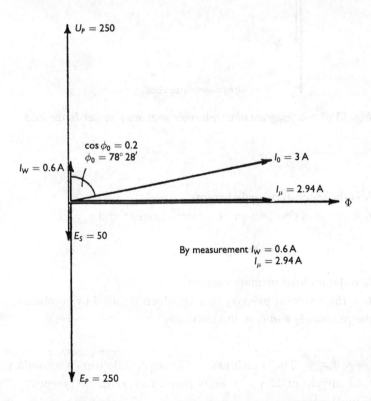

Fig. 62 Phasor diagram on no load (neglecting voltage drops)

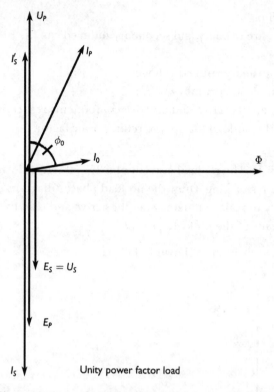

Fig. 63 Phasor diagram of transformer with unity power factor load

I_S is the secondary load current
I_S' is the load component of primary current and

$$I_S' = I_S \times N_S/N_P$$

I_0 is the no-load primary current
I_P is the resultant primary current which is found by combining the phasors I_S' and I_0 in the usual way.

EXAMPLE 4   The transformer of Example 3 delivers a secondary load current of 30 A at (i) unity power factor, (ii) 0.8 power factor lagging.

Draw the phasor diagram on load and determine the primary current in each case.

The load component of the primary current is

$$I_S' = I_S \times \frac{N_S}{N_P}$$

$$= 30 \times \frac{50}{250} \quad \text{(see Volume 2 for transformer ratio problems)}$$

$$= 6\,\text{A}$$

The phasor diagrams are shown in Figs 64 and 65.

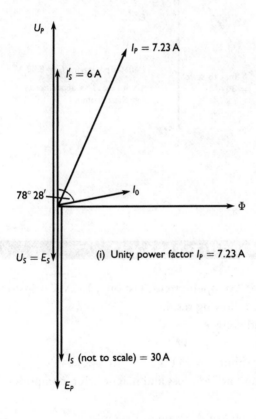

(i) Unity power factor $I_P = 7.23\,\text{A}$

Fig. 64

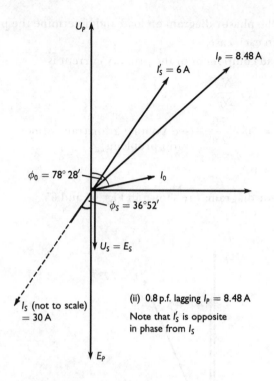

(ii) 0.8 p.f. lagging $I_P = 8.48$ A

Note that I'_S is opposite in phase from I_S

Fig. 65

OPEN-CIRCUIT TEST

EXAMPLE 5 An open-circuit test on a 15 kVA transformer yielded the following results:

applied voltage   230 V

current   3 A

power input   144 W.

Determine the iron-loss and magnetizing components of the no-load current.

From the phasor diagram of Example 3:

$$\frac{I_W}{I_0} = \cos \phi_0$$

where I_W is the iron-loss component of the open-circuit current I_0

and $\quad \cos\phi_0 = \dfrac{\text{power input}}{\text{VA input}}$

$$= \frac{144}{230 \times 3}$$

$\therefore \qquad \dfrac{I_W}{3} = \dfrac{144}{230 \times 3}$

$$I_W = \frac{144 \times 3}{230 \times 3}$$

$$= \underline{0.63\,\text{A}}$$

Also $\qquad \dfrac{I_\mu}{I_0} = \sin\phi_0$

where I_μ is the magnetizing component of I_0

$$\cos\phi_0 = \frac{144}{230 \times 3} = 0.2087$$

$$\phi_0 = 77°\,57' \quad \sin\phi_0 = 0.978$$

$$\frac{I_\mu}{3} = 0.978$$

$$I_\mu = 3 \times 0.978$$

$$= 2.934$$

$$= \underline{2.93\,\text{A}}$$

or using Pythagoras' theorem

$$I_0^2 = I_W^2 + I_\mu^2$$

and $\qquad I_\mu^2 = I_0^2 - I_W^2 \quad$ etc.

SHORT-CIRCUIT TEST AND EFFICIENCY

EXAMPLE 6 The transformer of Example 5 circulates full-load current on short-circuit when the power input is 200 W.

Calculate its efficiency at (a) full load, (b) half load when the power factor is unity in each case.

$$\text{efficiency} = \frac{\text{output}}{\text{input}}$$

$$= \frac{\text{output}}{\text{output} + \text{losses}}$$

(a) The output is 15 kVA at unity power factor = 15 kW.

The full-load losses

$$= \text{iron loss} + \text{copper loss at full load}$$

$$= 144 + 200$$

$$= 344\,\text{W}$$

efficiency $\eta = \dfrac{15\,000}{15\,000 + 344}$

$$= 0.978 \text{ or } 97.8\%$$

(b) Copper loss is proportional to current^2.

Thus copper loss at half load

$$= \left(\tfrac{1}{2}\right)^2 \times \text{copper loss at full load}$$

$$= \tfrac{1}{4} \times 200$$

$$= 50\,\text{W}$$

Iron losses are constant. Thus

$$\eta = \frac{\tfrac{1}{2} \times 15\,000}{\tfrac{1}{2} \times 15\,000 + 144 + 50}$$

$$= \frac{7500}{7500 + 194}$$

$$= 0.975 \text{ or } 97.5\%$$

EXAMPLE 7 Calculate the full-load efficiency of the transformer of Example 6 when the load power factor is 0.8 lagging.

$$\text{Output} = 15\,\text{kVA at } 0.8\,\text{p.f.}$$
$$= 15 \times 0.8$$
$$= 12\,\text{kW}$$
$$\eta = \frac{12\,000}{12\,000 + 344}$$
$$= \underline{97.2\%}$$

MAXIMUM EFFICIENCY

EXAMPLE 8 Calculate the value of the load for which the efficiency of the transformer considered will have its maximum value.

At maximum efficiency copper and iron losses are equal.

Let $1/n$ be the fraction of full load at which copper and iron losses are equal, then:

$$\left(\frac{1}{n}\right)^2 \times 200 = 144$$
$$\left(\frac{1}{n}\right)^2 = \frac{144}{200}$$
$$\frac{1}{n} = \sqrt{\frac{144}{200}}$$
$$= \sqrt{0.72}$$
$$= 0.8485$$

Thus maximum efficiency occurs when the load is

$$0.8485 \times 15 = \underline{12.73\,\text{kVA}}$$

ALL-DAY EFFICIENCY

All-day efficiency $\eta_D = \dfrac{\text{Energy output}}{\text{Energy input}}$ (over 24 hours)

EXAMPLE 9 The transformer previously considered is energized continuously but is on load of 15 kVA unity power factor for 15 hours only. Calculate its all-day efficiency.

Iron-loss energy over 24 hours $= 144 \times 24$

$$= 3456 \, \text{Wh} \quad \text{(watt-hour)}$$

Full-load copper-loss energy
over 15 hours $\qquad = 200 \times 15$

$$= 3000 \, \text{Wh}$$

Output energy $\qquad = 15\,000 \times 15$

$$= 225\,000 \, \text{Wh}$$

$$\eta_D = \frac{225\,000}{225\,000 + 3000 + 3456}$$

$$= \underline{0.972 \text{ or } 97.2\%}$$

EXAMPLE 10 A 1000 kVA industrial transformer has iron losses of 4700 W and copper losses of 12 000 W and is subject to the following daily load cycle throughout the year:

No load	6 hours per day
300 kVA	3 hours per day
600 kVA	9 hours per day
700 kVA	4 hours per day
900 kVA	2 hours per day.

Calculate the annual cost of the losses assuming that energy costs are 3.3 p per kWh.

Copper losses on no-load can be ignored so:

On 300 kVA copper loss $= 0.3^2 \times 12 \, \text{kW}$

$$= 1.08 \, \text{kW} \times 3 \, \text{hrs} = 3.24 \, \text{kWh}$$

On 600 kVA copper loss $= 0.6^2 \times 12 \, \text{kW}$

$$= 4.32 \, \text{kW} \times 9 \, \text{hrs} = 38.88 \, \text{kWh}$$

On 700 kVA copper loss $= 0.7^2 \times 12$ kW

$$= 5.88 \text{ kW} \times 4 \text{ hrs} = 23.52 \text{ kWh}$$

On 900 kVA copper loss $= 0.9^2 \times 12$ kW

$$= 9.72 \text{ kW} \times 2 \text{ hrs} = \underline{19.44 \text{ kWh}}$$

$$\text{Total daily copper loss} = 85.08 \text{ kWh}$$

$$\text{Daily iron loss} = 4.7 \text{ kW} \times 24 \text{ hrs} = \underline{112.8 \text{ kWh}}$$

$$\text{Total daily loss} = 197.88 \text{ kWh}$$

So annual operational loss $= 197.88 \times 365$

and annual cost of losses $= 197.88 \times 365 \times 3.3 = £2383.46$

REGULATION

Percentage regulation

$$= \frac{\text{change in secondary voltage from no load to full load}}{\text{open-circuit secondary voltage}} \times 100$$

EXAMPLE 11 Calculate the full-load terminal voltage of a transformer having 5% regulation and open-circuit secondary voltage 400 V.

Change in voltage $\quad\quad = \dfrac{5}{100} \times 400$

$$= 20 \text{ V}$$

Terminal voltage on full load $= 400 - 20$

$$= \underline{380 \text{ V}}$$

EXAMPLE 12 A transformer has a no-load and full-load secondary voltage of 500 V and 487 V respectively. Calculate the percentage regulation of the transformer.

$$\text{Percentage regulation} = \frac{U_s\,(\text{no-load}) - U_s\,(\text{on-load})}{U_s\,(\text{no-load})} \times 100$$

$$= \frac{500 - 487}{500} \times 100$$

$$= \frac{13}{500} \times 100$$

$$= 2.6\%$$

SYSTEM SHORT-CIRCUIT CONDITIONS

EXAMPLE 13 A factory generating station employs parallel connected three-phase generators and transformers as shown in Fig. 66. Calculate the short-circuit MVA and the short-circuit current in the event of a short circuit at switchgear positions:

A when switchgear B is in open condition and generator X operating;

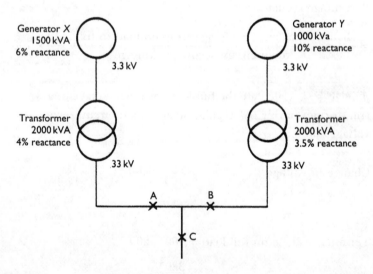

Fig. 66

B when switchgear A is in open condition and generator Y operating;

C when both switchgear A and B are in closed condition and both generators operating.

Ignore the impedance of the interconnecting cables.

Take a 1 MVA base.

L.H. generator $\dfrac{6 \times 1}{1.5} = 4\%$ reactance

L.H. transformer $\dfrac{4 \times 1}{2} = 2\%$ reactance

$\therefore$ Total reactance L.H. branch $= 4 + 2 = 6\%$

Short-circuit MVA at position A $= \dfrac{1 \times 100}{6}$

$$= 16.6 \, \text{MVA}$$

Short-circuit current at position A $= \dfrac{16.6 \times 10^6}{\sqrt{3} \times 33 \times 10^3}$

$$= \underline{290.4 \, \text{A}}$$

R.H. generator $\dfrac{10 \times 1}{1} = 10\%$ reactance

R.H. transformer $\dfrac{3.5 \times 1}{2} = 1.75\%$ reactance

$\therefore$ Total reactance R.H. branch $= 10 + 1.75 = 11.75\%$

Short-circuit MVA at position B $= \dfrac{1 \times 100}{11.75}$

$$= 8.5 \, \text{MVA}$$

Short-circuit current at position B $= \dfrac{8.5 \times 10^6}{\sqrt{3} \times 33 \times 10^6}$

$$= \underline{148.7 \, \text{A}}$$

Total reactance at position C $= \dfrac{1}{X} = \dfrac{1}{X_1} + \dfrac{1}{X_2}$

$$= \dfrac{1}{6} + \dfrac{1}{11.75} = 3.97\%$$

and short-circuit MVA $= \dfrac{1 \times 100}{3.97}$

$$= \underline{25.19\,\text{MVA}}$$

Short-circuit current at position C $= \dfrac{25.19 \times 10^6}{\sqrt{3} \times 33 \times 10^3}$

$$= \underline{440.7\,\text{A}}$$

EXERCISE 13

1. An 11 000/230 V transformer has 1500 primary turns. Calculate the number of secondary turns.
2. Calculate the primary current of a 6600/400 V transformer when its secondary current is 200 A.
3. A transformer for 50 Hz working has 1500 primary turns, 650 secondary turns, the cross-sectional area of the core is 1500 mm^2, and the maximum flux density is 0.75 T. Calculate the primary and secondary e.m.f.
4. A 230/12 V 50 Hz transformer has 2000 primary turns and the cross-sectional area of the core is 1000 mm^2. Calculate the maximum value of flux density in the core.
5. The primary supply voltage of a transformer is 500 V at 50 Hz. The cross-sectional area of the core is 2500 mm^2. Calculate the number of primary turns in order that the maximum value of flux density shall not exceed 0.75 T.
6. If the secondary voltage of the transformer of question 5 is to be 2000 V, calculate the number of secondary turns.
7. The magnetizing and iron-loss components of a 230/120 V transformer are respectively 5 A and 2 A. Draw the no-load phasor diagram accurately to scale and determine from it the no-load current and power factor.
8. The no-load current of a transformer is 4 A at 0.2 power factor lagging. Determine graphically or calculate the magnetizing and iron-loss components of the no-load current.
9. An open-circuit test on a 5 kVA transformer yielded the following results:

Primary voltage 230 V; primary current 2 A; power input 84.4 W.

Calculate the magnetizing and iron-loss components of the no-load current.

10. The magnetizing and loss components of the no-load current of an 11 000/240 V transformer are respectively 2.5 A and 0.5 A.

    By means of an accurately constructed phasor diagram or otherwise determine the primary current when the secondary current is 100 A at:

    (a) unity power factor;

    (b) 0.8 power factor lagging;

    (c) 0.8 power factor leading.

11. Calculate the efficiency of the transformer of question 9 for a range of values of load at unity power factor up to full load. Plot a graph of efficiency against the fraction of full load. State from the graph the value of load at which the transformer has its maximum efficiency and calculate the copper loss at this value of load.

12. (a) Explain the differences between a single-phase double-wound transformer and a single-phase auto-transformer.

    What are the requirements of BS 7671 regarding the use of auto-transformers?

    (b) A 40 kVA single-phase transformer was tested for efficiency by the 'open-circuit' and 'short-circuit' tests. On short circuit, at full-load current, the power used was 1140 W. On open circuit, the power used was 800 W.

    Calculate the efficiency of the transformer at unity power factor on (i) full load, and (ii) half full load.   (C & G)

13. (a) Why is it usually necessary to cool transformers? Describe briefly two methods by which this can be done.

    (b) A 20 kVA transformer when tested was found to have 600 watts iron losses, and 700 watts copper losses when supplying full load at unity power factor.

    Calculate the efficiency of the transformer at unity power factor (i) on full load, (ii) on half load.   (C & G)

14. The full-load copper and iron losses of a 50 kVA transformer are respectively 250 W and 150 W. Calculate the efficiency of the transformer on (i) full load, (ii) half load, when the load power factor is 0.8 lagging in each case.

15. **(a)** Explain why transformers need to be cooled.

   (b) A 600 kVA three-phase transformer is immersed in a tank containing 2 m^3 of insulating oil. The efficiency of the transformer at full load is 97%.

   Calculate the average rise in temperature in °C of the oil after a 3 hour run at full load and unity power factor, assuming that 60% of the heat energy lost in the transformer is expended in heating the oil.

   Specific heat of oil = 2135 J/(kg °C).

   1 m^3 of oil weighs 900 kg.   (C & G)

16. A 25 kVA transformer has iron loss 500 W and full-load copper loss 650 W. Determine the value of unity power factor load at which the efficiency will have its maximum value and calculate that value.

17. A 20 kVA transformer operates with maximum efficiency of 98% when on 0.9 of its full load at unity power factor. Calculate its efficiency at full load.

18. A 15 kVA transformer has iron loss 100 W and full-load copper loss 125 W. It is energized continuously and supplies the following loads over a 24 hour period:

   from 0800 to 1800 hours 8 kVA at unity power factor;

   from 1600 to 2400 hours 6 kVA at 0.8 power factor lagging.

   Calculate the all-day efficiency.

19. A 100 kVA transformer has ordinary efficiency 98% on full-load unity power factor when its iron and copper losses may be taken as equal. Over a period of 24 hours it supplies loads as follows:

   0000 to 1200 hours 40 kW at unity power factor;

   0200 to 2200 hours 35 kVA at 0.75 power factor lagging.

   Calculate its all-day efficiency.

20. The open-circuit voltage of a transformer is 415 V. On full load, the terminal voltage is 400 V. Calculate the percentage regulation.

21. A 230 V transformer has 5% regulation. Calculate its terminal voltage on full load.

22. Using Fig. 66 from example 13, the transformer on the left is replaced by a new 2000 kVA transformer having a reactance of 5%. Calculate the new short-circuit MVA and short-circuit current at switchgear position A and position C.

Electrostatics

References: **Electric field.**
Field strength.
Flux density.
Parallel plate capacitor.
Series and parallel arrangement of capacitors.
Quantity of charge and energy stored.
Charging and discharging curves.

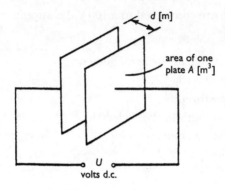

Fig. 67

Figure 67 shows an arrangement of two parallel metal plates each having cross-sectional area A (m^2) and separated by insulation of thickness d (m) is called a capacitor. When it is connected to a d.c. supply of U volts it becomes charged. The following facts are known:

1. The quantity of charge in coulombs is

    $Q = CU$ where C is the capacitance (farad)

2. The electric field strength

    $E = U/d$ (volts/metre)

3. The electric flux density

$$D = Q/A \quad (\text{coulomb/metre}^2)$$

4. The energy stored

$$W = \tfrac{1}{2}CU^2 \quad (\text{joule})$$

5. The capacitance of the arrangement is

$$C = \varepsilon_0 \varepsilon_r \frac{A}{d} \quad (\text{farad})$$

where ε_0 is the permittivity of free space with numerical value 8.85×10^{-12}; ε_r is the relative permittivity of the insulator or dielectric.

EXAMPLE 1 Two parallel metal plates each of area $0.01\ \text{m}^2$ and separated by a layer of mica 2 mm thick and of relative permittivity 6 are connected to a 100 V d.c. supply.

Calculate

(a) the capacitance of the arrangement;
(b) the charge stored;
(c) the energy stored;
(d) the field strength in the dielectric;
(e) the electric flux density.

(a) $\quad C = \varepsilon_0 \varepsilon_r \dfrac{A}{d}$

$$= 8.85 \times 10^{-12} \times 6 \times 0.01 \times \frac{1}{2/1000}$$

$$= 265.5 \times 10^{-12}\ \text{farad}$$

$$= \underline{265.5}\ (\text{pF})\ \text{picofarad}$$

$$(1\,\text{F} = 10^6\,\mu\text{F} = 10^{12}\,\text{pF})$$

(b) $\quad Q = CU$

$$= 265.5 \times 10^{-12} \times 100$$

$$= 26\,550 \times 10^{-12}\ \text{coulomb}$$

$$= 0.02655 \times 10^{-6}\ \text{coulomb}$$

$$= \underline{0.02655}\ \text{microcoulomb}\ (\mu\text{C})$$

(c) $W = \frac{1}{2}CU^2$

$= \frac{1}{2} \times 265.5 \times 100^2 \times 10^{-12}$

$= 132.75 \times 10^{-8}$ joule

$= 1.3276 \times 10^{-6}$ joule

$= \underline{1.3275}$ microjoule (μJ)

(d) $E = \dfrac{U}{d}$

$= \dfrac{100}{2/1000}$

$= \underline{50\,000}$ volts/metre

(e) $D = Q/A$

$= \dfrac{0.02655 \times 10^{-6}}{0.01}$

$= \underline{0.02655 \times 10^{-4}}$ coulomb/metre^2

SERIES ARRANGEMENT OF CAPACITORS

If a number of capacitors of values C_1, C_2, C_3, etc. are connected in series (see Fig. 68) they are equivalent to a single capacitor of value C given by

$$\frac{1}{C} = \frac{1}{C_1} + \frac{1}{C_2} + \frac{1}{C_3} \quad \text{etc.}$$

When the arrangement is connected to a d.c. supply of U volts the charge stored is the same on each and equal to $Q = CU$.

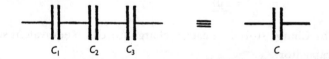

Fig. 68

EXAMPLE 2 Calculate the value of a capacitor which when connected in series with another of $20\,\mu F$ will give a resulting capacitance of $12\,\mu F$.

$$\frac{1}{C} = \frac{1}{C_1} + \frac{1}{C_2}$$

$$\frac{1}{12} = \frac{1}{20} + \frac{1}{C_2}$$

$$\frac{1}{C_2} = \frac{1}{12} - \frac{1}{20}$$

$$\frac{1}{C_2} = \frac{5-3}{60}$$

$$= \frac{2}{60}$$

$$C_2 = \frac{60}{2} = 30\,\mu\text{F}$$

The required value is thus $30\,\mu\text{F}$.

EXAMPLE 3 Capacitors of $4\,\mu\text{F}$, $6\,\mu\text{F}$ and $12\,\mu\text{F}$ are connected in series to a 300 V d.c. supply. Calculate:

(a) the equivalent single capacitor;

(b) the charge stored on each capacitor;

(c) the p.d. across each capacitor;

(d) the energy stored in each capacitor.

(a)
$$\frac{1}{C} = \frac{1}{4} + \frac{1}{6} + \frac{1}{12}$$

$$= \frac{3+2+1}{12}$$

$$= \frac{6}{12}$$

$$C = \frac{12}{6} = 2\,\mu\text{F}$$

(b) Charge stored on each = charge stored on equivalent single capacitor

$$Q = CU$$

$$= 2 \times 300$$

$$= 600\,\mu\text{C} \quad \text{(microcoulomb because capacitance}$$
$$\text{is in microfarad)}$$

(c) The p.d. on each capacitor is found by using the formula $Q = CU$ where C is the appropriate value of capacitance and Q is as calculated above.

Rearranging the formula gives

$$U = \frac{Q}{C}$$

Thus for the $4\,\mu F$ capacitor

$$U_4 = \frac{600}{4} = \underline{150\,V}$$

Similarly

$$U_6 = \frac{600}{6} = \underline{100\,V}$$

and $\quad U_{12} = \frac{600}{12} = \underline{50\,V}$

(Note that these sum to 300 V.)

(d) The energy stored is calculated by using the formula $W = \frac{1}{2}CU^2$ with the appropriate capacitance and voltage calculated above.

Thus for the $4\,\mu F$ capacitor

$$W_4 = \frac{1}{2} \times \frac{4}{10^6} \times (150)^2$$

$$= \underline{0.045\,J}$$

$$W_6 = \frac{1}{2} \times \frac{6}{10^6} \times (100)^2$$

$$= \underline{0.03\,J}$$

and $\quad W_{12} = \frac{1}{2} \times \frac{12}{10^6} \times (50)^2$

$$= \underline{0.015\,J}$$

PARALLEL ARRANGEMENT OF CAPACITORS

If a number of capacitors of values C_1, C_2, C_3, etc. are connected in parallel (see Fig. 69) they are equivalent to a single capacitor of

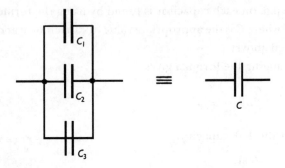

Fig. 69

value C given by

$$C = C_1 + C_2 + C_3 \quad \text{etc.}$$

When the arrangement is connected to a d.c. supply of U volts the total charge is the sum of the charges stored separately on each, i.e. if Q be the total charge

$$Q = Q_1 + Q_2 + Q_3$$

where Q_1 is the charge on C_1, etc. and $Q_1 = C_1 U$, etc. The voltage U is common to all the capacitors.

EXAMPLE 4 Capacitors of $4\,\mu\text{F}$ and $5\,\mu\text{F}$ are connected in parallel and charged to 20 V. Calculate the charge stored on each and the total stored energy.

The p.d. is the same on each capacitor. The charge on the $4\,\mu\text{F}$ capacitor is

$$Q_4 = C_4 U$$
$$= 4 \times 20$$
$$= \underline{80\,\mu\text{C}} \quad \text{(microcoulombs because C is in microfarads)}$$

Similarly

$$Q_5 = C_5 U$$
$$= 5 \times 20$$
$$= \underline{100\,\mu\text{C}}$$

The total energy may be calculated either by finding the energy stored separately on each capacitor and summing or by considering the total capacitance thus:

$$C = 4 + 5$$

$$= 9\,\mu F$$

and total energy

$$W = \tfrac{1}{2}CU^2$$

$$= \tfrac{1}{2} \times 9 \times 20^2$$

$$= \underline{1800\,\mu J} \quad \text{(microjoules because } C \text{ is in microfarads)}$$

EXAMPLE 5   Calculate the value of a single capacitor equivalent to the arrangement shown in Fig. 70.

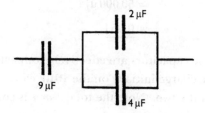

Fig. 70

The capacitor equivalent to the parallel group

$$= 2 + 4$$

$$= 6\,\mu F$$

The circuit then reduces to that shown in Fig. 71

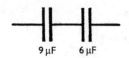

Fig. 71

153

and the equivalent capacitor of value C is given by

$$\frac{1}{C} = \frac{1}{9} + \frac{1}{6}$$

$$= \frac{2+3}{18}$$

$$C = \frac{18}{5} = \underline{3.6\,\mu F}$$

EXAMPLE 6   A capacitor of $10\,\mu F$ is fully charged from a $100\,V$ d.c. supply and immediately connected to an uncharged capacitor of $15\,\mu F$. Calculate (a) the energy stored initially in the $10\,\mu F$ capacitor and (b) the total energy stored subsequently in the two capacitors.

Initial energy stored $W_1 = \frac{1}{2}CU^2$

$$= \frac{1}{2} \times 10 \times (100)^2$$

$$= 50\,000\,\mu J$$

$$= \underline{0.05\,J} \qquad\qquad (a)$$

Subsequently the capacitors are effectively connected in parallel and the charge initially on the $10\,\mu F$ capacitor is then shared between the two. (Note the total *charge* is constant, the *energy* is not.)

Charge stored on the $10\,\mu F$ capacitor

$$Q = CU$$

$$= 10 \times 100$$

$$= 1000\,\mu C$$

The two capacitors in parallel are equivalent to a single one of value $10 + 15 = 25\,\mu F$.

The common p.d. on the two capacitors is found by using the formula $Q = CU$. Thus

$$1000\,\mu C = 25\,\mu F \times U$$

$$U = 40\,V$$

The final energy stored

$$W_2 = \tfrac{1}{2} \times 25 \times (40)^2$$

$$= 20\,000\,\mu J$$

$$= \underline{0.02\,J} \qquad \qquad \text{(b)}$$

The difference between (a) and (b) is the energy dissipated during the sharing of the charge between the two capacitors.

THE MULTIPLATE CAPACITOR

An arrangement of n parallel plates each of effective area A (m^2) and separated by insulation of thickness d (m) and of relative permittivity ε_r produces $(n - 1)$ capacitors in parallel and the capacitance of the arrangement is (see Fig. 72)

$$C = (n - 1)\varepsilon_0 \varepsilon_r \frac{A}{d} \quad F$$

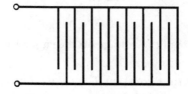

Fig. 72

EXAMPLE 7 A parallel plate capacitor has 21 semicircular plates each of diameter 5 cm. The plates are 2 mm apart in air. Calculate the capacitance of the arrangement.

$$C = (n - 1)\varepsilon_0 \varepsilon_r \frac{A}{d}$$

$$= (21 - 1) \times \frac{8.85}{10^{12}} \times 1 \times \frac{1}{2} \times \frac{\pi}{4} \times \frac{5^2}{10^4} \times \frac{1}{2/1000} \quad F$$

Notes (i) The relative permittivity of air is taken to be 1.
(ii) The formula for the area of a circle of diameter d is $\pi d^2/4$; the area of a semicircle is one-half of this.

(iii) The conversions of square centimetres to square metres and millimetres to metres.

$$C = 20 \times \frac{8.85}{10^{12}} \times \frac{1}{2} \times \frac{\pi}{4} \times \frac{5^2}{10^4} \times \frac{1000}{2} \times 10^{12} \, \text{pF} = \underline{86.9 \, \text{pF}}$$

D.C. EXCITED CIRCUIT HAVING RESISTANCE AND CAPACITANCE IN SERIES

If such a circuit having capacitance C farad in series with resistance R ohms is supplied at U volts d.c. (Fig. 73), the voltage on the capacitor at any instant t seconds after closing the switch S is found as follows:

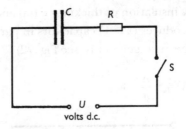

Fig. 73

(1) The final value of the voltage on the capacitor is U, the supply voltage.

(2) Calculate the time constant $T = CR$ (seconds).

(3) Draw graph axes to suitable scales as in Fig. 74 (the time required for the voltage to reach its maximum value may be taken as five times the time constant.)

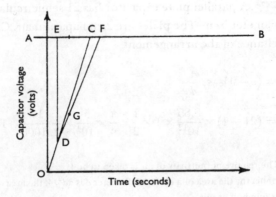

Fig. 74

(4) Draw the horizontal line AB so that OA = U. Mark AC = T. Join OC. Select any point D on OC near to O. Project upwards vertically from D to E on AB. Mark EF = T. Join FD. Repeat the procedure for a new point G on DF and so on until the complete curve is traced.

During the charging of the capacitor the current decays in a manner illustrated graphically in Fig. 75.

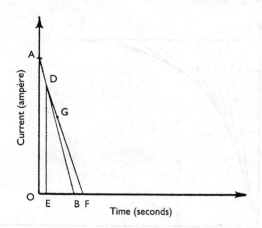

Fig. 75

1. Calculate the initial value of the current

   $$I = U/R$$

2. The time constant is as before

   $$T = CR$$

3. Draw graph axes to suitable scales.

4. Mark OA = I and OB = T. Join AB and select any point D on AB close to A. Project from D to E on the time axis and mark EF = T. Join DF. Select a new point G on DF and repeat the procedure until a complete curve is traced.

When a capacitor of value C farad has been fully charged to U volts and is then discharged through a resistor of R ohms, curves showing the decay of both capacitor voltage and current are constructed in exactly the same way as that described above.

EXAMPLE 8 A capacitor of $20\,\mu\text{F}$ and a resistor of $5\,\text{M}\Omega$ are connected in series to a d.c. supply of $100\,\text{V}$. Determine graphically, for the instant when the switch has been closed for 3 minutes,

(a) the p.d. on the capacitor;

(b) the charge and energy stored in the capacitor;

(c) the charging current.

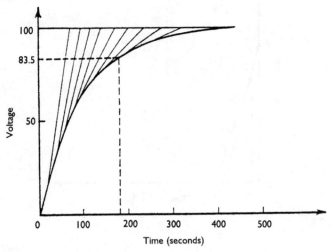

Fig. 76

For (a) and (b) the curve shown in Fig. 76 is required.

Time constant $T = CR$

$$= 5 \times 10^6 \times \frac{20}{10^6}$$

$$= 100 \text{ seconds}$$

From the curve at $t = 180$ seconds

    Capacitor voltage $U_c = \underline{83.5\,\text{V}}$ (a)

The charge

$$Q = CU_c$$

$$= \frac{20}{10^6} \times 83.5$$

$$= 167\,\mu\text{C} \quad \text{(microcoulomb)}$$

The energy

$$W = \frac{1}{2}CU_c^2$$

$$= \frac{1}{2} \times \frac{20}{10^6} \times (83.5)^2$$

$$= \underline{0.0697\,\text{J}} \qquad\qquad\qquad\qquad\qquad\qquad (b)$$

(c) may be determined by constructing the curve showing the decay of current in the manner described or by using Kirchhoff's law for the voltages acting in the circuit, for the p.d. across the resistor is $i \times R$ where i is the current at the instant considered, and

$$U = U_c + i \times R$$

$$100 = 83.5 + i \times 5 \times 10^6$$

$$i \times 5 \times 10^6 = 100 - 83.5$$

$$= 16.5$$

$$i = \frac{16.5}{5 \times 10^6}\,\text{A}$$

$$= \frac{16.5 \times 10^6}{5 \times 10^6}\,\mu\text{A} \quad \text{(micro ampère)}$$

$$i = \underline{3.3\,\mu\text{A}} \qquad\qquad\qquad\qquad\qquad\qquad (c)$$

EXERCISE 14

$(\varepsilon_0 = 8.85 \times 10^{-12})$

1. A capacitor consists of two parallel metal plates each 100 mm by 120 mm and separated by a sheet of insulation having relative permittivity 7.5 and of thickness 1.5 mm. Calculate:
   (a) its capacitance;
   (b) the charge and energy stored when the capacitor is charged to 75 V;
   (c) the electric flux density and voltage gradient under these conditions.

2. Calculate the diameter of circular plates required to produce a capacitance of 100 pF if the plates are separated by 0.5 mm of insulation of relative permittivity 5.

3. A capacitor of 100 pF is charged to a p.d. of 100 V. Calculate the charge and energy stored. Calculate also the energy stored if the distance between the plates is (a) halved, (b) doubled. (Remember that the charge (Q) stored is constant; both C and U vary.)

4. Capacitors of 3 μF and 5 μF are connected in series to a 240 V d.c. supply. Calculate:
   (a) the resultant capacitance;
   (b) the charge on each;
   (c) the p.d. on each;
   (d) the energy stored in each.

5. Calculate the value of a single capacitor equivalent to three 24 μF capacitors connected in series. What would be the value of 10 such capacitors connected in series?

6. What value of capacitor connected in series with one of 20 μF will produce a resultant capacitance of 15 μF?

7. Three capacitors of values 8 μF, 12 μF and 16 μF respectively are connected across a 240 V d.c. supply (i) in series, and (ii) in parallel. Calculate in each case, the resultant capacitance, and also the potential difference across each capacitor.   (C & G part question)

8. Calculate the value of the single capacitor equivalent to the arrangement shown in Fig. 77.

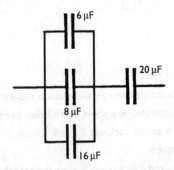

Fig. 77

9. Capacitors of 12 μF and 20 μF are connected in parallel. A third capacitor of 64 μF is connected in series with these two in parallel and the whole circuit is supplied with 300 V d.c. Calculate the charge and the energy stored in each capacitor.

10. A capacitor of 60 µF is fully charged from a 200 V d.c. supply and then connected to an uncharged capacitor of 40 µF. Calculate:

    (a) the charge and energy stored initially in the 60 µF capacitor;

    (b) the potential difference and the energy stored in the combination. Account for the difference in the values of energy stored.

11. A multiplate capacitor consists of 25 plates each of area 600 mm^2. The plates are separated in air by a distance of 0.5 mm. Calculate the capacitance of the capacitor.

12. The 12 plates of a multiplate capacitor are semicircular and 40 mm in diameter. They are fixed so that adjacent plates are 0.75 mm apart in air. Calculate the capacitance of the arrangement.

13. A 60 µF capacitor is charged from a 100 V d.c. supply through a 0.5 MΩ resistor. Construct a curve showing:

    (a) charging current plotted against time;

    (b) the capacitor voltage plotted against time.

14. A 10 µF capacitor having been charged to 340 V is discharged through a 5 MΩ resistor. Construct a curve showing the capacitor voltage plotted against time.

15. The following results were obtained when a capacitor of unknown value was charged to 150 V and then discharged through a 5 MΩ resistor. If the time constant (CR) is equal to the time required, from the commencement of the discharge, for the current in the circuit to fall to 0.368 of its maximum value, estimate the value of the capacitor.

Time after discharge commences (s)	10	20	30	50	60	70	80
current (µA)		23.6	18.6	15.0	9.5	7.5	6.2	5.0

Time after discharge commences (s)	90	100	120	130	150	200	250
current (µA)		4.0	3.0	2.0	1.5	0.7	0.3	0

16. A 15 µF capacitor is charged to a potential difference of 200 V and then discharged through a 50 kΩ resistor.

    (a) Calculate (i) the time constant; (ii) the initial value of the discharge current.

    (b) Construct a graph showing the decay of the discharge current with respect to the time from switching on.

    (c) From the graph find the value of the current after a period of 1.5 s.

Utilization of electric power II
Illumination

References: **Illumination terms and measurement.**
Point-by-point lighting calculation using the inverse
square law and cosine law.
Utilization factor (UF).
Light loss factor (LLF).
Lumen method.

INVERSE SQUARE LAW

A lamp of luminous intensity I candela in all directions below the
horizontal when suspended d metres above a surface (see Fig. 78),
produces illumination at P below the lamp given by

$$E_P = \frac{I}{d^2} \text{ lumen per square metre or lux (lx)}$$

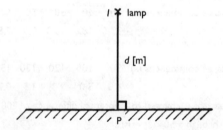

Fig. 78

COSINE LAW

The illumination at any other point Q, x metres from P and on the
same horizontal plane through P (see Fig. 79) is given by

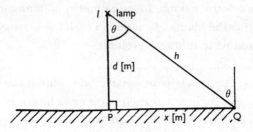

Fig. 79

$$E_Q = \frac{I}{h^2} \cos \theta \, \text{lx}$$

where h and θ are as shown in Fig. 79.

Also $\qquad h^2 = d^2 + x^2$ (Pythagoras)

and $\qquad \cos \theta = \dfrac{d}{h}$

$$\qquad\qquad\quad = \frac{d}{\sqrt{d^2 + x^2}}$$

EXAMPLE I A luminaire producing luminous intensity 1500 candela in all directions below the horizontal is suspended 4 m above the floor. Calculate the illumination produced at a point P immediately below the luminaire and at a point Q 2.5 m away from P.

The situation is as illustrated in Fig. 79.

The illumination at P

$$E_P = \frac{1500}{4^2}$$

$$\quad = \underline{93.75 \, \text{lx}}$$

The illumination at Q

$$E_Q = \frac{1500}{22.25} \times \frac{4}{4.717}$$

$\qquad\qquad$ (h and d being calculated as shown above)

$$\quad = \underline{57.2 \, \text{lx}}$$

The illumination at a point due to a number of luminaires is found by calculating the illumination due to each luminaire separately and then adding to find the resultant.

EXAMPLE 2   An area is 16 m square and is illuminated by four luminaires, one mounted at each corner at a height of 6 m. The luminaires each have luminous intensity 1000 cd in all directions below the horizontal. Calculate the illumination produced at the centre of the square.

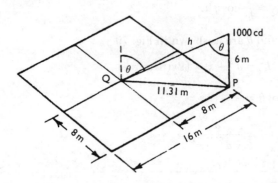

Fig. 80

Figure 80 shows the situation as far as a single luminaire is concerned. In order to calculate $\cos \phi$ it is first necessary to determine the length PQ where Q is at the centre of the square.

By Pythagoras $PQ^2 = 8^2 + 8^2$

$$= 128$$

and $PQ = 11.31 \text{ m}$

as before $\cos \theta = \dfrac{6}{\sqrt{6^2 + 11.31^2}}$

$$= \dfrac{6}{12.81}$$

also $h^2 = 6^2 + 11.31^2$

$$= 164$$

The illumination due to one luminaire

$$= \frac{1000}{164} \times \frac{6}{12.81}$$

$$= 2.856\,\text{lx}$$

The total illumination

$$= 4 \times 2.856$$

$$= \underline{11.42\,\text{lx}}$$

LUMINOUS-FLUX CALCULATIONS USING UTILIZATION (UF) AND LIGHT LOSS (LLF) FACTORS

$$\text{Total luminous flux} = \frac{\text{service value of illuminance} \times \text{area}}{\text{utilization factor} \times \text{light loss factor}}$$

where the illuminance is given in lux and the area is measured in square metres.

The utilization factor takes account of the fact that not all of the luminous flux emitted from the lamp or luminaire actually falls into the working area, e.g. when light is lost by dark windows or non-reflective dark decorative surfaces.

The light loss factor takes account of the reduced illumination provided by an averagely dirty lighting installation compared with that provided by the same installation when it is clean. It is generally the anticipated level of illumination when a new lighting installation has been in service for some time.

EXAMPLE 3 Estimate the total luminous flux required to provide a service value of 120 lx in a room 5 m by 7 m. Utilization and light loss factors are respectively 0.6 and 0.8.

Substituting the values in the formula:

$$\text{Total luminous flux} = \frac{120 \times 5 \times 7}{0.6 \times 0.8}$$

$$= \underline{8750\,\text{lm}}$$

EXAMPLE 4 Calculate the total power required for the installation of Example 3 if the luminaires used have an efficacy of 12 lumens per watt.

$$\text{Power} = \frac{\text{total lumens}}{\text{lumens per watt}}$$

$$= \frac{8750 \; \cancel{\text{lm}}}{12 \; \cancel{\text{lm}}/\text{W}}$$

$$= \underline{729 \, \text{W}}$$

SPACING–HEIGHT RATIO

For a regular square arrangement of fittings this is

$$\frac{\text{distance between adjacent luminaires}}{\text{height of luminaires above the working plane}}$$

EXAMPLE 5 Estimate a suitable spacing between luminaires which have a spacing height ratio of 1.5 and are suspended 4 m above the working plane.

If S is the spacing and H the height

$$\frac{S}{H} = 1.5$$

$$S = 1.5 \times 4$$

$$= \underline{6 \, \text{m}}$$

EXAMPLE 6 An open-plan office building, 50 m long and 20 m wide, is to be illuminated using 100 twin-tube fluorescent luminaires. Each tube has an output of 8000 lumens and the input to each luminaire is 300 W.

Determine

(a) the average illuminance if the light loss factor is 0.9 and the utilization factor is 0.6;

(b) the efficacy of the luminaire;

(c) the minimum current rating of the main control gear for the lighting installation if the luminaires are arranged with

5 luminaires to a circuit and balanced as near as is practicable, the power factor being 0.85 and the supply being at 400/230 V 50 Hz.

(a) Average illuminance $= \dfrac{100 \times 2 \times 8000}{50 \times 20 \times 0.9 \times 0.6}$

$= 2963\,\text{lx}$

(b) Efficacy $= \dfrac{2 \times 8000}{300}$

$= 53.33\,\text{lm/W}$

(c) Number of circuits $= \dfrac{100}{5}$

$= 20$ circuits

Assume circuits are arranged:

R phase, 7 circuits; Y phase, 7 circuits; B phase, 6 circuits

Minimum control gear current rating

$= \dfrac{300 \times 5 \times 7}{230 \times 0.85}$

$= 53.7\,\text{A}$ (63 A practicable current rating)

EXAMPLE 7   A hall, 15 m by 20 m, is to be illuminated to a level of 70 lx. Luminaires having an efficacy of 12 lm/W and spacing–height ratio 1.2 are to be suspended 4 m above the floor. Estimate the number of luminaires required and the power of each luminaire. Assume a utilization factor 0.5 and a light loss factor of 0.8.

Total lumens required from all luminaires

$$= \frac{\text{lux} \times \text{area}}{\text{utilization factor} \times \text{light loss factor}}$$

$$= \frac{70 \times 15 \times 20}{0.5 \times 0.8}$$

$$= \underline{52\,500\,\text{lm}}$$

Next find the distance between adjacent luminaires:

$$\frac{S}{H} = 1.2$$

$$\therefore \quad S = 1.2 \times 4$$

$$= \underline{4.8\,\text{m}}$$

The number of rows of luminaires

$$= \frac{\text{width of room}}{\text{spacing}}$$

$$= \frac{15\,\text{m}}{4.8\,\text{m}}$$

$$= \underline{3\,\text{say}}$$

The number of luminaires per row

$$= \frac{\text{length of room}}{\text{spacing}}$$

$$= \frac{20}{4.8}$$

$$= \underline{4\,\text{say}}$$

Total number of luminaires

$$= 4 \times 3$$

$$= \underline{12}$$

Lumens per luminaire

$$= \frac{52\,500}{12}$$

$$= \underline{4375\,\text{lm}}$$

Watts per luminaire

$$= \frac{4375\,\cancel{\text{lm}}}{12\,\cancel{\text{lm}}/\text{W}}$$

$$= \underline{364.6\,\text{W}}$$

300 W luminaires will probably suffice.

A scale plan of the room can be drawn to show the positions of the luminaires (Fig. 81).

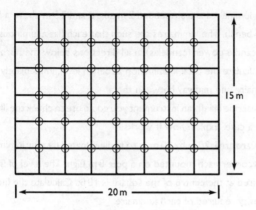

Fig. 81

1. A square area 10 m by 10 m is illuminated by four luminaires. Each
   luminaire is mounted on a pole 7.5 m high at one corner of the square and
   emits 1050 cd in all directions. Determine the level of illumination at points:
   (a) at the centre of the square;
   (b) at the foot of one pole;
   (c) midway along one of the sides of the square.

2. An advertisement board, 5 m square, is fixed to a wall with the bottom
   edge near the ground. A luminaire, giving a luminous intensity of 4000
   candelas in all directions towards the board, is fixed level with the bottom
   of the board, and 6 m distant, giving maximum illumination at the centre of
   the bottom edge.
   Calculate the illumination:
   (a) at the centre of the bottom edge;
   (b) at the centre of the top edge;
   (c) at one of the top corners.
   Suggest a method of giving reasonably even illumination over the whole
   board.   (C & G)

3. A luminaire is suspended 2 m above a level workbench, such that the
   luminous intensity in all directions below the horizontal is 400 candelas.
   Calculate the illumination at a point A on the surface of the bench
   immediately below the luminaire, and at other bench positions 1 m, 2 m
   and 3 m from A in a straight line.   (C & G)

4. Two luminaires are suspended 2 m apart, and 2.5 m above a level workbench. The luminaires are such that each has a luminous intensity of 200 candle-power (candelas) in all directions below the horizontal.

Calculate the total illumination at bench level, immediately below each luminaire and midway between them.

Describe briefly an instrument you could use to check the illumination in such a case. Explain how it works. (C & G)

5. A square area, 20 m by 20 m, is to be illuminated by four luminaires, one at each corner each mounted on a pole 8 m high. The level of illumination required at the centre of the square is 10 lx. Calculate the luminous intensity required of each luminaire.

6. A circular assembly hall 15 m in diameter is to be illuminated to a general level of 200 lx. Utilization and light loss factors may be taken as 0.6 and 0.8 respectively. Estimate the power required to illuminate the hall:

(a) using tungsten luminaires having an efficacy of 14 lm/W;

(b) using fluorescent luminaires having an efficacy of 40 lm/W.

7. A workshop is 9 m by 20 m. An illumination of 200 lx is required and this is to be obtained by using luminaires with an efficacy of 40 lm/W. The utilization factor may be taken as 0.45 and a light loss factor of 0.7 should be allowed. Calculate:

(a) the power required;

(b) the cost of the energy used in a month of 20 days if the charge is on the basis of 8p per unit for the first 60 units and 5p for each additional unit, and the lighting is in use 4 hours per day.

8. A room 7 m by 10 m is to be used as a general office and must be provided with illumination to a level of 400 lx. 150 W luminaires are to be installed giving a utilization factor 0.5 and requiring a light loss factor of 0.8. Assuming that the efficacy of the luminaires is 13 lm/W, calculate the number of luminaires required.

9. Describe one form of portable instrument for measuring values of illumination in different parts of a room.

An office, 30 m long by 15 m wide, is to be illuminated to an intensity of 400 lx. Assuming the average lumen output of the luminaires is 30 per watt, the utilization factor 0.5, and the light loss factor 0.8, calculate the total wattage required. (C & G)

10. A church hall is 40 m long and 14 m wide. It is to be illuminated by a number of luminaires suspended 3 m above the floor. The luminaires have a spacing–height ratio of 1.75. Estimate the number of luminaires required and draw a scale plan of the building marking the positions of the luminaires.

11. Draw a scale plan of an office 20 m long and 6 m wide to show the number and positions of the lighting points if the luminaires are mounted 4 m above the floor and have a spacing–height ratio of 1.8.

12. A light assembly shop, 17 m long, 8 m wide and 3 m to trusses, is to be illuminated to a level of 200 lx. The utilization and light loss factors are respectively 0.54 and 0.8. Make a scale drawing of the plan of the shop and set out the required lighting points, assuming the use of tungsten luminaires. You may assume an efficacy of 13 lm/W. (C & G)

13. A room 28 m by 7 m is provided with sixteen 300 W luminaires having an efficacy of 13 lm/W. Assuming a utilization factor of 0.48 and a light loss factor of 0.8, determine the general level of illumination provided by the luminaires.

14. A machine shop 50 m by 27 m is illuminated to a level of 300 lx by incandescent luminaires which have an efficacy 16 lm/W. It is decided to replace these luminaires by fluorescent luminaires having an efficacy of 40 lm/W so that the overall level of illumination remains the same. Estimate the saving in the cost of electrical energy per 8-hour shift assuming:
    (a) that all the luminaires are in use during the 8 hours;
    (b) utilization and light loss factors are the same for both types of luminaires at 0.5 and 0.8 respectively;
    (c) electrical energy costs 7p per unit.

15. A warehouse is 30 m by 20 m. Illumination is to be provided by a number of 1000 W luminaires having an efficacy of 16 lm/W and a spacing–height ratio of 1.5. The luminaires are suspended 4 m above the working surface. The utilization factor is 0.53 and the light loss factor is 0.75.
    (a) Estimate the number of luminaires required.
    (b) Determine the general level of illumination produced.

16. (i) Explain the points that should be considered when planning the electric lighting of one of the following:
    (a) a workshop with rows of benches for small assembly work;
    (b) a large drawing office.
    (ii) An office 20 m by 50 m needs an average illumination at desk level of 400 lx. The following alternatives are available:

(a) 80 W fluorescent luminaires emitting 2800 lumens when new;

(b) 150 W tungsten filament luminaires emitting 13 lm/W when new.

Calculate the number of luminaires needed for each alternative assuming the utilization factor for the room to be 0.6 and the light loss factor to be 0.85. (C & G)

17. A workshop measuring 15 m by 25 m by 3.5 m high, used for simple bench fitting of small parts, needs a general illumination at bench level of 400 lx. The following schemes are suggested:

(a) 80 W fluorescent luminaires emitting 40 lm/W;

(b) 200 W tungsten filament luminaires emitting 13 lm/W.

Assuming the utilization factor to be 0.65 and the light loss factor to be 0.8 calculate the number of luminaires to be installed for each scheme.

For each case draw a diagram giving your suggested layout of the luminaires showing their spacing and mounting height. (C & G)

Utilization of electric power III Heating schemes

References: **Power and energy.**
Specific heat capacity.
Heat loss.

EXAMPLE I A storage heater contains $0.1\,\text{m}^3$ of water. The 230 V heating element produces a temperature rise of 85 °C in $1\frac{1}{2}$ hours and the efficiency of the device is 82%. Calculate the rating of the heater in watts and the resistance of its element. The specific heat of water is 4187 joules per kilogram per degree Celsius (4187 J/kg °C).

In a heating operation

Energy in the form of heat = mass × temperature rise

×specific heat

$$Q = m \times (\theta_2 - \theta_1) \times c$$

where m is the mass of substance in kg

c is its specific heat in J/(kg °C)

θ_2 and θ_1 are the upper and lower temperatures in °C

1 m^3 of water weighs 10^3 kg.

Substituting the given values:

$$Q = 0.1 \times 10^3 \, \text{kg} \times 85 \, \text{degC} \times 4187 \, \frac{\text{J}}{\text{kg degC}}$$

$$= 0.1 \times 10^3 \times 85 \times 4187 \, \text{J}$$

and the energy to be supplied is

$$W = 0.1 \times 10^3 \times 85 \times 4187 \times \frac{100}{82}$$

(allowing for inefficiency)

$$= 43\,402\,000 \, \text{J}$$

This energy is supplied over a period of $1\frac{1}{2}$ hours, that is,

$1\frac{1}{2} \times 3600$ seconds

Energy supplied per second

$$= \frac{43\,402\,000 \, \text{J}}{1.5 \times 3600 \, \text{s}}$$

$$= 8037 \, \text{W}$$

or 8.037 kW

which is the power rating of the element.

The relationship between power and resistance is

$$P = \frac{U^2}{R}$$

$$8037 = \frac{230^2}{R}$$

$$R = \frac{230^2}{8037}$$

$$= \underline{6.58\,\Omega}$$

which is the resistance of the element.

EXAMPLE 2 A certain process in a factory requires a steel tank holding 1000 litres of cleaning fluid. This is to be raised from 14 °C to 40 °C in 4 hours by single-phase industrial-type immersion heaters connected to a 400/230 V 50 Hz supply, assuming a heat transfer efficiency of 90%. (Specific heat of the cleaning fluid is 4000 J/kg °C.)

The control panel for the heaters is sited 53 m from its BS 88 part 2 fuses in a distribution board, the ambient temperature in the area is 45 °C and the four cables between the distribution board and control panel are to be single-core p.v.c. insulated drawn into steel conduit. Maximum volts drop in circuit is limited to 6 volts.

Establish the:

(a) heat input to the cleaning fluid;
(b) power rating of the heaters;
(c) current demand;
(d) rating of the BS 88 Part 2 fuses;
(e) minimum cable rating;
(f) maximum mV/A/m value;
(g) minimum cable c.s.a.;
(h) voltage drop in the cables.

(a) Heat input to fluid

$$= 1000 \times 4000 \times (40 - 14)$$

$$= 104\,\text{MJ}$$

(b) Power input to heaters

$$= \frac{104 \times 10^6 \times 100}{4 \times 60 \times 60 \times 90}$$

$$= 8\,\text{kW}$$

(c) Current demand

$$= \frac{8000}{\sqrt{3} \times 400}$$

$$= 11.6\,\text{A}$$

(d) Rating of fuses (I_n): 16 A.

(e) Minimum cable rating (I_t) using BS 7671 Table 4C:

$$= \frac{16}{0.87}$$

$$= 18.4\,\text{A}$$

(f) Maximum mV/A/m value

$$= \frac{6 \times 1000}{11.6 \times 53}$$

$$= 9.76\,\text{mV/A/m}$$

(g) From BS 7671 Appendix 4 Tables 9D1A and 4D1B, minimum cable c.s.a. is $4\,\text{mm}^2$ (24 A: 9.5 mV/A/m).

(h) Voltage drop in cables

$$= \frac{11.6 \times 53 \times 9.5}{1000}$$

$$= 5.84\,\text{V}$$

EXAMPLE 3   A room has dimensions 4 m by 6 m by 2.5 m. Electric heaters are to be provided to produce an average temperature rise of 8 °C. Calculate the rating of the heaters required assuming two changes of air occur per hour and that 40 % of their output is wasted.

The density of air is $1.28\,\text{kg/m}^3$ and its specific heat is $1000\,\text{J/(kg\,°C)}$.

Volume of room $= 4 \times 6 \times 2.5$

$$= 60\,\text{m}^3$$

Mass of air in the room $= 60\,\cancel{\text{m}} \times 1.28\dfrac{\text{kg}}{\cancel{\text{m}}} = 76.8\,\text{kg}$

Heat content of the air for each hour

$$Q = 2 \times 76.8\,\text{kg} \times 8\,^\circ\text{C} \times 1000\,\frac{\text{J}}{\text{kg}\,^\circ\text{C}}$$

$$= 2 \times 76.8 \times 8 \times 1000\,\text{J}$$

Electrical energy required per hour

$$W = 2 \times 76.8 \times 8 \times 1000 \times \frac{100}{60}$$

(only 60% of the energy is usefully employed)

$$= 20.5 \times 10^5 \text{ joules per hour}$$

$$= \frac{20.5 \times 10^5}{3600}\,\text{J/s}$$

$$= \underline{570\,\text{W}}$$

EXAMPLE 4 A building is 6.5 m by 8 m and has ceiling height 3 m. The windows have total area $7\,\text{m}^2$ and there are wooden doors of total area $3.5\,\text{m}^2$. Allowing two complete changes of air per hour, calculate the power required to maintain a 20 °C temperature difference between the inside and outside of the building. Heat transmission coefficients in watts per square metre per degree Celsius are:

Brick (walls)	1.88
Concrete (floor)	1.13
Plaster (ceiling)	2.84
Wood	3.98
Glass	5.4

The density of air is $1.28\,\text{kg/m}^3$ and its specific heat is $1000\,\text{J/(kg}\,^\circ\text{C)}$.

The calculation is performed in two sections:

(i) To calculate the heat losses through walls, etc.

$$\text{Area of brick wall} = \text{total area} - \text{area of doors}$$
$$- \text{area of windows}$$

$$= (2 \times 6.5 \times 3) + (2 \times 8 \times 3)$$

$$- 3.7 - 7$$

$$= 76.5\,\text{m}^2$$

Let P_b be the power wasted in the form of heat through the brick:

$$P_b = 1.88 \frac{W}{m^2 \, ^\circ C} \times 76.5 \, m^2 \times 20 \, ^\circ C$$

$$= 1.88 \times 76.5 \times 20$$

$$= 2877 \, W$$

Similarly for the wood

$$P_W = 3.98 \times 3.5 \times 20$$

$$= 278.6 \, W$$

and for the glass

$$P_G = 5.4 \times 7 \times 20$$

$$= 756 \, W$$

and for the concrete

$$P_C = 1.13 \times 8 \times 6.5 \times 20$$

$$= 1175 \, W$$

and for the plaster

$$P_P = 2.84 \times 8 \times 6.5 \times 20$$

$$= 2954 \, W$$

Total power loss $= 2877 + 278.6 + 756 + 1175 + 2954$

$$= 8041 \, W$$

(ii) To calculate the heat required to warm the air:

mass of air within the room

$$= \text{volume} \times \text{density}$$

$$= 6.5 \times 8 \times 3 \, m^3 \times 1.28 \, \frac{kg}{m^3}$$

$$= 199.68 \, kg$$

Heat required to raise the temperature of twice this mass of air through 20 °C

$$= \text{mass} \times \text{temperature rise} \times \text{specific heat}$$

$$= 199.68\,\text{kg} \times 20\,°\text{C} \times 1000\frac{\text{J}}{\text{kg}\,°\text{C}} \times 2$$

$$= 7\,987\,200 \text{ joules per hour}$$

$$= \frac{7\,987\,200}{3600}\text{joules per second}$$

$$= 2219\,\text{W}$$

Total power required $= 2219 + 8041$

$$= 10\,260\,\text{W}$$

$$\underline{\text{or } 10.26\,\text{kW}}$$

EXAMPLE 5 In a large building, the used air is passed through a heat pump. The waste heat is recovered and used to heat water.
(a) The waste heat energy recovered from the air is 72×10^6 joules per hour and the thermal efficiency of the system is 65%. Determine the quantity of water that may be heated from 15 °C to 50 °C in one hour. (Specific heat capacity of water is 4180 J/kg °C.)
(b) At peak water demand, 500 litres per hour are required. Any shortfall is to be made up by means of an immersion heater with an efficiency of 85%. Determine the rating of the immersion heater required.
(c) The motor for the heat pump has an output of 5 kW and an efficiency of 75%. Calculate the gain (ratio of input to output energy) of the system at peak water demand.　　　　(C & G)
(a) Heat delivered to water in one hour

$$= 72 \times 10^6 \times 4180 \times (50 - 15)$$

Quantity of water heated from 15 °C to 50 °C in one hour

$$= \frac{72 \times 10^6 \times 0.65}{4180 \times 35}$$
$$= 320 \text{ litres}$$

(b) Additional water to be heated

$$= 500 - 320$$

$$= 180 \text{ litres}$$

Immersion heater rating required

$$= \frac{180 \times 4180 \times 35}{3600 \times 0.85}$$

$$= 8.6 \text{ kW}$$

(c) Input to motor $= \dfrac{5.0}{0.75}$ $\qquad = 6.67 \text{ kW}$

Input to immersion heater $\qquad = 8.6 \text{ kW}$

Total input $= 6.67 + 8.6$ $\qquad = 15.27 \text{ kW}$

Output to water $= \dfrac{500 \times 4180 \times 35}{3600} = 20.3 \text{ kW}$

Gain of system $= \dfrac{20.3}{15.27}$ $\qquad = 1.33$

EXAMPLE 6 A works reception office $12 \text{ m} \times 10 \text{ m}$ with a ceiling height of 4 m has a door area 7 m^2 and a window area of 25 m^2. The office is heated electrically to maintain an average inside temperature of $18\,^{\circ}\text{C}$ when the outside temperature is $0\,^{\circ}\text{C}$. Assuming that there are two changes of air per hour, calculate the total kW rating of the heating system.

Specific heat capacity of air	$1010 \text{ J/kg}\,^{\circ}\text{C}$
Density of air	1.292 kg/m^3
Heat transfer coefficients	(U values)
Walls – brick and plaster	$1.7 \text{ W/m}^2\,^{\circ}\text{C}$
Floor – timber on concrete	$0.8 \text{ W/m}^2\,^{\circ}\text{C}$
Ceiling – plaster	$1.2 \text{ W/m}^2\,^{\circ}\text{C}$
Doors – wood	$2.8 \text{ W/m}^2\,^{\circ}\text{C}$
Windows – double glazed	$2.9 \text{ W/m}^2\,^{\circ}\text{C}$

(C & G)

Employing the 'per hour' basis:

Volume of air heated $= 2 \times 12 \times 10 \times 4\,\mathrm{m}^3/\mathrm{hour}$

Mass of air heated $= 960 \times 1.29\,\mathrm{kg/hour}$

$$= 1240\,\mathrm{kg/hour}$$

Quantity of heat energy Q required to raise the temperature of air from $0\,^\circ\mathrm{C}$ to $18\,^\circ\mathrm{C}$:

$$Q/\mathrm{hour} = mc(\phi_2 - \phi_1)\,\mathrm{J/hour}$$

$$= 1240 \times 1010(18 - 0)\,\mathrm{J/hour}$$

$$= 22.54\,\mathrm{MJ/hour}$$

Now $1\,\mathrm{kWh} = 3.6\,\mathrm{MJ}$.

Thus electrical energy required per hour

$$= \frac{22.54\,\mathrm{kWh}}{3.6}$$

$$= 6.26\,\mathrm{kWh}$$

Heat Losses

Area of brickwork

$$= (2 \times \text{length} \times \text{height}) + (2 \times \text{breadth} \times \text{height})$$

$$-\text{window and door area}$$

$$= (2 \times 12 \times 4) + (2 \times 10 \times 4) - (7 + 25)\,\mathrm{m}^2$$

$$= (96 + 80 - 32)\,\mathrm{m}^2$$

$$= 144\,\mathrm{m}^2$$

Area of floor and ceiling $= 12 \times 10 = 120\,\mathrm{m}^2$

Power loss/$^\circ$C difference through walls

$$= 144 \times 1.7 = 244.8\,\mathrm{W}/^\circ\mathrm{C\ diff.}$$

Power loss/$^\circ$C difference through floor

$$= 120 \times 0.8 = 96\,\mathrm{W}/^\circ\mathrm{C\ diff.}$$

Power loss/$^\circ$C difference through ceiling

$$= 120 \times 1.2 = 144\,\mathrm{W}/^\circ\mathrm{C\ diff.}$$

Power loss/°C difference through doors

$$= 7 \times 2.8 = 19.6 \, \text{W/°C diff.}$$

Power loss/°C difference through windows

$$= 25 \times 2.9 = 72.5 \, \text{W/°C diff.}$$

Total power loss/°C difference

$$= (244.8 + 96 + 144 + 19.6 + 72.5)$$
$$= 576.9 \, \text{W/°C difference}$$

Thus power needed to cover heat loss from office

$$P = (576.9 \times 18) \, \text{W} = 10.4 \, \text{kW}$$

and total power needed to heat office

$$P = (10.4 + 6.26) \, \text{kW} = 16.66 \, \text{kW}$$

EXAMPLE 7 A sports centre swimming pool holds 60 000 litres of water and it is to be initially heated by an electrode boiler connected to a 'time-of-day' tariff supply available between midnight and 8.30 a.m. daily.

The temperature of the water is to be raised by the boiler from 10 °C to 20 °C in 72 hours.

It is anticipated that heat will be lost from the water at the rate of 20 MJ per hour and this loss will be made up by a heat pump also connected to the above tariff supply and having a 'gain' of three. (Specific heat of water is 4180 J/kg °C.)

(a) Calculate the minimum power rating of the electrode boiler, neglecting the hourly heat losses.

(b) Calculate the minimum power input to the heat pump.

(a) Heat input $= 60\,000 \times 4180 \times (20 - 10)$

$$= 2508 \, \text{MJ}$$

Operating time of electrode boiler $= 8.5 \times 60 \times 60 \times 3$

$$= 91\,800 \, \text{s}$$

Power input to boiler $= \dfrac{2508 \times 10^6}{91\,800} = 27.3 \, \text{kW}$

(b) Losses per day $= 20\,000 \times 10^3 \times 24$

$$= 480 \times 10^6 \, \text{J}$$

provided by the heat pump in $8\frac{1}{2}$ hours. So

$$P = \frac{480 \times 10^6}{8.5 \times 60 \times 60}$$

$$= 15.69 \, \text{kW} \, (15.7 \, \text{kW})$$

Since $\text{gain} = \dfrac{\text{Output}}{\text{Input}}$

$$\text{Input to heat pump} = \frac{15.7}{3}$$

$$= 5.23 \, \text{kW}$$

EXERCISE 16

Specific heat of water 4187 J/(kg °C)
Specific heat of air 1000 J/(kg °C)
Density of air 1.28 kg/m³

1. With the aid of sketches, describe the operation of a thermostatically controlled immersion heater as fitted into domestic hot water systems.

    A 230 V storage heater contains 0.055 m³ of water. The temperature of the water is raised from 18 °C to 88 °C in $1\frac{1}{4}$ hours with an efficiency of 88%. Assuming the current remains constant throughout the run, find the resistance of the element during the operation, the number of units used and the nominal size of the heater in kW.

2. Describe with clear sketches a free-outlet electric water heater as installed over a wash basin or kitchen sink, and explain how it works.

    A heater of this type rated at 750 W holds 0.007 m³ of water. If it takes 58 minutes to raise the temperature of the water from 20 °C to 90 °C, calculate the efficiency of the water heater.

3. A domestic hot-water cylinder holds 0.136 m³ of water and is fitted with a 3 kW heater. Assuming an efficiency of 70%, calculate the time required to raise the temperature of the water from 5 °C to 70 °C.

4. An industrial process requires 0.2 m³ per hour of liquid of specific heat 2000 J/(kg °C) and density 900 kg/m³ to be raised in temperature by 80 °C. Assuming that 35% of the heat supplied is wasted, calculate the rating of the

heater in kW required for the process. Determine also the resistance of the element for operation from a 230 V supply.

5. Write a short account of the design factors to be taken into account when considering the electrical heating of a room.

   A room 11 m by 6 m by 3.5 m high is heated by means of low-temperature tubular heaters with a total loading of 9 kW. Assume that 40% of the heat supplied is used to heat the air in the room.

   Calculate the average temperature rise of the air if there are two complete changes of air per hour.

6. Describe with sketches the construction and operation of a domestic-type thermal storage room heater.

   A room 5 m by 3.5 m by 3 m high is heated by a thermal storage heater. The heater takes in electrical energy during the night from 2200 hrs to 0700 hrs and releases the stored heat during the day from 0700 hrs to 2200 hrs. The average temperature rise in the room is 20 °C and there are two complete changes of air in each hour. Assuming that 45% of the energy supplied to the heater is used to heat the air in the room, calculate (a) the kWh input to the heater in 24 hours, (b) the kW rating of the heater.

7. A room is 5 m by 4 m and is 3 m high. It is required to maintain the temperature within the room at 9 °C above that of the surroundings and there are to be two complete changes of air per hour. Calculate the rating of the heaters required assuming that the system is 65% efficient.

8. A workshop 13 m by 6.5 m with ceiling height 3.5 m has a window area of 35 m² and a door area of 10 m². The workshop is to be heated electrically so that the average temperature inside is maintained at 18 °C when the temperature outside is 0 °C.

   Using the information given in Example 4, calculate the power required in kW on the assumption that there will be two complete changes of air per hour.

9. A glass house on a concrete base is 6.5 m long. The end section is that of a rectangle 2.5 m wide and 1.5 m high surmounted by a triangle 0.75 m high. Using the information given in Example 4 and assuming $1\frac{1}{2}$ complete changes of air per hour, calculate the rating of the heaters necessary to maintain the inside of the building at a temperature 17 °C above that of the outside.

10. A brick building with concrete floor and plaster ceiling is 25 m long, 10 m wide and 3.5 m high. There are windows of total area 36 m² and wooden doors of total area 16 m². It is required to maintain the temperature inside the building 18 °C above that of the outside with two changes of air per hour by using electric heaters. Use the information given in Example 4 to determine the rating of the heaters required.

Answers

Exercise I

1. (a) $10 = 12I_1 - 6I_2$
            $0 = -6I_1 + 15I_2$
  (b) $0 = 16I_1 - 6I_2$
            $-10 = -6I_1 + 11I_2$

  (c) $6 = 13I_1 - 4I_2$
            $-8 = -4I_1 + 16I_2$
  (d) $-2 = 17I_1 - 2I_2$
            $4 = -2I_1 + 17I_2$

  (e) $0 = 10I_1 - 5I_2 - 4I_3$
            $0 = -5I_1 + 10I_2 - 3I_3$
            $2 = -4I_1 - 3I_2 + 8I_3$
  (f) $2 = 7I_1 - 4I_2$
            $0 = -4I_1 + 22I_2 - 6I_3$
            $-2 = -6I_2 + 18I_3$

2. (a) $I_{BA} = I_{AF} = I_{FE} = \frac{1}{7}$ A; $I_{BC} = I_{CD} = I_{BE} = \frac{2}{21}$ A; $I_{EB} = \frac{5}{21}$ A
  (b) $I_{BA} = I_{AF} = I_{FE} = 0.4$ A; $I_{DE} = I_{CD} = I_{BC} = 0.2$ A; $I_{EB} = 0.6$ A
  (c) $I_{BA} = I_{AH} = I_{HG} = 3.428$ A; $I_{CB} = I_{GF} = 1.875$ A; $I_{DC} = I_{FE} = I_{ED} = 5.938$ A;
     $I_{GB} = 1.563$ A; $I_{CF} = 4.063$ A

3. $I_{BD} = \frac{1}{24}$ A   4. $I_{10} = 0.563$ A; 5.63 V

5. Current through battery of internal resistance $1\,\Omega = 0.25$ A
   Current through battery of internal resistance $2\,\Omega = 0.125$ A
   Total charging current $= 0.375$ A

6.

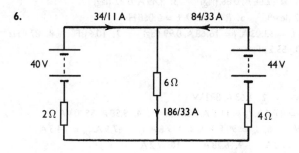

Fig. 82

7. 3 mA, D to B   8. (a) $49\,\Omega$, 6.428 A, (b) $25\,\Omega$

9. $I_{AB} = 240$ A; $I_{BC} = 180$ A; $I_{CD} = 100$ A; $U_B = 225$ V;
   $U_C = 214$ V; $U_D = 210$ V

10. $I_{AB} = 158.3$ A; $I_{BC} = 98.3$ A; $I_{CD} = 18.3$ A; $I_{DA} = 81.7$ A; $U_B = 227$ V;
   $U_C = 221$ V; $U_D = 220$ V

11. $I_{PQ} = 77.5$ A; $I_{QR} = 7.5$ A; $I_{RP} = -42.5$ A; $U_O = 231$ V; $U_R = 230$ V

12.

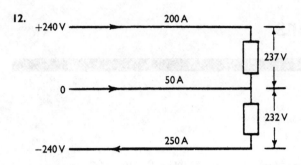

Fig. 83

Exercise 2

1. (a) 18.7 Ω, (b) 12.3 A, (c) 0.8 lead, (d), U_R 184.5 V; U_L 125.75 V; U_C 261 V
2. 1.523 A
3.

Inductance (H)	0.3	0.253	1.5	1	0.0135
Capacitance (F)	25	40	1.69	50	30
Resonant frequency (Hz)	58	50	100	33	250

4. (a) 5.94 A, (b) U_R 143 V, U_L 299 V, U_C 118 V, (c) 31.4 Hz, 9.3 A
5. (a) 18.64 A, 2904 W, 0.648 lag, (b) 169 μF   **6.** 24 μF, 0.714 lead; 425 μF, 0.714 lag
7. (a) 0.318 H, (b) 48.9 μF, (c) 99 V   **8.** 100 Hz, 34.7 Hz
9. (a) 26.5 Hz, (b) 25 A, (c) $U_L = U_C = 2080$ V; $U_R = 100$ V

Exercise 3

1. 1.35 A (lag)   **2.** 2.66 A; 0.86 (lag)   **3.** 3.49 A; 0.92 (lag)
4. 0.609 A; 0.992 (lead)   **5.** $R = 163$ Ω; $L = 0.068$ H
6. $I_C = 14.38$ A, $I_L = 13.03$ A, $I = 16.68$ A, 0.99 (lag)   **7.** 10.6 μF   **8.** 87.4 Hz
9. 1.64 A   **10.** 55.2 μF

Exercise 4

1. 4.62 A, 2561 W   **2.** 1.63 A, 882 W
3. $I_R = 9.2$ A, $I_Y = 17.7$ A, $I_B = 11.5$ A, 5221 W   **4.** 9.58 A, 5520 W
5. 6.9 A, 5060 W   **6.** $I_R = 99.8$ A, $I_Y = 119.8$ A, $I_B = 47.9$ A, $I_N = 64.3$ A
7. 52.9 A   **8.** 1.68 A   **9.** 6.28 A   **10.** 7.2 A
11. (a) (i) 18.61 Ω, (ii) 12.36 A, (iii) 0.54 (lag), (iv) 57.50°
  (b) (i) 26.54 Ω, (ii) 8.66 A
  (c) (i) 6.9 A, (ii) 0.97 (lag)

Exercise 5

1. (a) 7.7 A, 7.7 A, 5520 W, (b) 13.25 A, 23 A, 16 560 W   **2.** 19.21 A, 7388 W
3. (a) 5.15 A, 3732 W, (b) 2.98 A, 1248 W
4. (i) 2.25 A, 0.469 (lag), 759.4 W, (ii) 6.75 A, 0.469 (lag), 2278 W
5. (a) 6.9 Ω, (b) 20.8 Ω   **6.** (a) 884 μF, (b) 295 μF

7. $I_{RY} = 6.63\,\text{A}$, $I_{YB} = 7.95\,\text{A}$, $I_{BR} = 12.5\,\text{A}$, 5390 W

8. (a) 16.48 A, (b) 28.56 A, (c) 20 029 W

Exercise 6

1. 22 kW, 5 kVAr (lag), 22.6 kVA, 0.974 (lag)   **2.** 0.885 (lag), 99.45 A

3. (a) 36.2 A, (b) 29.25 A (c) 0.87 (lag)

4. 30.1 kW, 21.3 kVAr, 36.9 kVA, 0.82 (lag), 53.2 A

5. (a) 6.68 kVAr (lead), (b) (i) 123 µF, (ii) 41.2 µF

6. (a) 89 kVA, (b) 0.807 (lag), (c) 71.7 kW, (d) 387 A

7. 198 A, 0.956 (lag)   **8.** 147 µF, 441 µF   **9.** (a) 148 µF (b) synchronous motor

10.

	kVA	kW	kVAr	p.f.	Line current
a	15	12	9 (lag)	0.8 (lag)	21.65 A
b	12	12	0	1.0	17.32 A
c	8	0	8 (lead)	0	11.55 A
d	14.4	11.5	8.63 (lag)	0.8 (lag)	20.78 A
Overall values	36.8	35.5	9.63 (lag)	1.0	53.1 A

11. (a) 49 kW, (b) 0.97. (Hint: $W_1 + W_2 = \sqrt{3}U_L I_L \cos\phi$; $W_1 - W_2 = U_L I_L \sin\phi$; $\sin\phi/\cos\phi = \tan\phi = \sqrt{3}(W_1 - W_2)/(W_1 + W_2)$, hence ϕ and $\cos\phi$.)
(c) 50.5 kVA, (d) 72.9 A

12. 8.11p   **13.** 29.6 kW   **14.** (a) £37 501, 6.25p, (b) £39 205, 6.53p

Exercise 7

1. 385 V, 3.6%, 756 W   **2.** (a) 410 V, (b) 1160 W   **3.** 12.14 V   **4.** 95 mm²

5. 467 mm², 500 mm², 4.61 V   **6.** 5.8 V   **7.** 70 mm²   **8.** 25 mm²

9. 70 mm²   **10.** 14 A

11. (a) 17.32 A, (b) 20 A, (c) 19.2 mV/A/m, (d) 29.35 A, (e) 6 mm², (f) 3.79 V, (g) 25 mm

12. (a) (i) 1210 A, (ii) 0.25 s, (iii) 0.393 Ω, (iv) 585.2 A, (v) 3 s; (b) 0.44 Ω

13. (a) (i) 62.36 A, (ii) 63 A, (iii) 3.2 mV/A/m, (iv) 67.02 A, (v) 16 mm², (vi) 4.49 V, (b) (i) 0.51 Ω, (ii) 0.86 Ω (Table 41D)

14. (a) 96.51 A, (b) 100 A, (c) (i) 97.08 A, (ii) 70 mm², (iii) 1.82 V

15. (a) 36.08 A, (b) 40 A, (c) 42.55 A, (d) 2.92 mV/A/m, (e) 16 mm², (f) 8.22 V

16. (a) 17.32 A, (b) 20 A, (c) 9.6 mV/A/m, (d) 21.3 A, (e) 4 mm², (f) 4.94 V, (g) 13.3 A (satisfactory), (h) Table 5C factor = 225, Table 5D factor = 260, satisfactory

Exercise 8

1.

Flux density B (T)	1.2	1.3	0.8	500	0.45
Cross-sectional area A (m²)	0.5	0.006	0.65	0.002	0.035
Total flux ϕ (Wb)	600 mWb	0.78 mWb	520 mWb	1000 mWb	15.75 mWb

2. 0.000 442 Wb   **3.** 0.8 T   **4.** 0.8 T   **5.** 420 At/m   **6.** 700 At

7. 500   **8.** 0.551 A   **9.** (a) 0.0251 T, (b) 0.132 T, (c) 0.943 T

11. 227 000 At/m   **12.** 4.86 A   **13.** about 1 A   **14.** about 4 A

15. about 0.35 mWb   **16.** 0.397 A   **17.** 0.767 A   **18.** 4.61 mWb

19. 0.462 mWb   **20.** 905 At   **21.** 0.8 A

Exercise 9

3. 10 H   **4.** 1.42 ms   **5.** 0.03 J, 0.005 J, 500 J, 0.0002 J   **6.** 240 J   **7.** 150 Ω

8. (a) 108 H, (b) 1667 J, (c) 90 Ω, (d) 833.5 J

Exercise 10

1. (a) 82.5 A, (b) 259 V, (c) 1961 W   **2.** 213 V   **3.** 0.036 Wb

4. 12.63 rev/s   **5.** 12.35 rev/s   **6.** 255 V   **7.** (a) 6.94 kW, (b) 69.8 Nm

8. (a) 2 A, (b) 30 A, (c) 243 V   **9.** (a) 73.46%, (b) 1084 W   **10.** 234 V, 238.4 V

11. (a) 105.67 A, (b) 254.12 V, (c) field 360 W, armature 1340 W   **12.** 25 A

13. (a) 75 A, (b) 244.13 V, (c) 3.09 A, (d) 78.09 A, (e) 249.13 V

Exercise 11

1. 190 Nm   **2.** 23.83 rev/s   **3.** 19.88 rev/s   **4.** 1410 Nm   **5.** 17 A

6. 13.33 rev/s   **7.** 45.71 rev/s   **8.** 7.78 kW, 74%, 113 Nm

9. 16.78 A, 215 V, 63.13%, 31.9 Nm   **10.** 9.87 rev/s   **11.** 23.7 rev/s

12. 86.4%   **13.** 4147 W, 71.5%   **14.** 82.1%   **15.** 4.88 Ω

16. 1.71 Ω   **17.** 1.96 Ω   **18.** 6.3 Ω

Exercise 12

1. 1648 W   **2.** 73.9 Nm   **3.** 57.67 Nm   **4.** 110.3 Nm   **5.** 89.2 A

6. 2.26 A   **7.** 78%, 0.692 (lag)   **8.** 12.7 rev/min, 2575 Nm

9. 5.269 kW, 16.4 A   **10.** 24.7 A, 130 Nm

Exercise 13

1. 31.36 (31 T)   **2.** 12.1 A   **3.** 375 V, 162.5 V   **4.** 0.541 T   **5.** 1200

6. 4800   **7.** 5.39 A, 0.372 (lag)   **8.** $I_w = 0.8$ A, $I_\mu = 3.92$ A

9. $I_\mu = 1.97$ A, $I_w = 0.36$ A   **10.** (a) 3.68 A, (b) 4.42 A, (c) 2.54 A

12. (a) 95.4%, (b) 94.9%   **13.** (a) 93.9%, (b) 92.8%   **14.** (a) 99%, (b) 98.9%

15. 30.35 °C   **16.** 21.9 kW, 95.6%   **17.** 98.04%   **18.** 83.9%

19. 96.7%   **20.** 3.614%   **21.** 228 V

22. (A) 15.38 MVA, 269 A, (C) 23.9 MVA, 418 A

Exercise 14

1. (a) 53.1 pF, (b) 3983 pC, 0.149 µJ, (c) 0.332 µC/m^2, 50 kV/m   **2.** 83.78 mm

3. 0.01 C, 0.5 µJ, (a) 0.25 µJ, (b) 1 µJ

4. (a) 1.88 µF, (b) 450 µC, (c) 150 V, 90 V, (d) 0.034 J, 0.02 J

5. 8 µF, 2.4 µF   **6.** 60 µF

7. (a) 3.7 μF, 111 V, 74 V, 56 V, (b) 36 μF, 240 V   8. 12 μF
9. $Q_{64} = 6400$ μC, $W_{64} = 0.32$ J
   $Q_{12} = 2400$ μC, $W_{12} = 0.24$ J
   $Q_{20} = 4000$ μC, $W_{20} = 0.4$ J
10. (a) 12 000 μC, 1.2 J, (b) 120 V, 0.72 J   11. 255 pF   12. 81.52 pF
15. 8.6 μF   16. (a) (i) 0.75 s, (ii) 0.004 A (4 mA), (c) 0.42 mA

Exercise 15

1. (a) 28.8 lx, (b) 28.7 lx, (c) 28 lx   2. (a) 111 lx, (b) 50.4 lx, (c) 47.2 lx
3. 100 lx, 71.6 lx, 35.4 lx, 17.1 lx   4. 47.2 lx, 51.2 lx
5. 1340 cd   6. (a) 5.26 kW, (b) 1.84 kW   7. (a) 2.857 kW, (b) £13.23
8. 36   9. 15 kW   12. 24   13. 123 lx   14. £21.27
15. (a) 15, (b) 159 lx   16. (a) 280, (b) 402   17. (a) 90, (b) 110

Exercise 16

1. 13 Ω, 5.1, 4 kW   2. 98.48%   3. 4 hours 53 minutes   4. 12.3 kW, 4.3 Ω
5. 22 °C   6. (a) 24.9 kWh, (b) 2.77 kW   7. 591 W   8. 17.2 kW
9. £21.27   10. 28.6 kW